**Native Orchids
of the Southern
Appalachian
Mountains**

Native Orchids of the Southern Appalachian Mountains

Stanley L. Bentley

The University of North Carolina Press

Chapel Hill and London

© 2000 The University of North Carolina Press
All rights reserved

Designed by April Leidig-Higgins
Set in Electra by Eric M. Brooks
Printed in Hong Kong by C & C Offset Printing Co., Ltd.

The paper in this book meets the guidelines for permanence and durability of the Committee on Production Guidelines for Book Longevity of the Council on Library Resources.

Library of Congress Cataloging-in-Publication Data
Bentley, Stanley L. Native orchids of the southern Appalachian Mountains / by Stanley L. Bentley.
p. cm. Includes bibliographical references (p.) and index.
ISBN 0-8078-2563-8 (cloth: alk. paper) —
ISBN 0-8078-4872-7 (pbk.: alk. paper)
1. Orchids — Appalachian Region, Southern — Identification.
2. Orchids — Appalachian Region, Southern — Pictorial works.
I. Title.
QK495.O64 B47 2000 584'.4'0975 — dc21 99-087675

04 03 02 01 00 5 4 3 2 1

This book is dedicated to my wife, Mary Ann. She has thus far tolerated me for 35 years, a feat that, on its own, is no small accomplishment.

Mary Ann has traveled with me through oppressive heat in coastal swamps and bone-chilling cold in the Canadian Rockies. She has stood steadfast through violent thunderstorms while holding cover over my camera equipment and has plodded through thigh-deep, boggy mud. Once she pulled me from waist-deep muck—likely saving my life in the doing. She has experienced bee stings, swarms of ravenous mosquitoes, scores of chigger bites, and dozens of black fly bites. She has accidentally grabbed a snake, and she has nudged me back when I got too close to a black bear. All this was done in the search for orchids.

But Mary Ann has also enjoyed with me the beauty of nature throughout the southern Appalachian Mountains and all across North America. She has seen every orchid that I have seen. On the few occasions when she was not along as I found an orchid new to me, the joy has never been complete until I have been able to return to the site and share my discovery with her.

This book, then, is for Mary Ann.

One of these days, someone will give us a handbook of our wildflowers, by the aid of which we shall all be able to name those we gather without the trouble of analyzing them.
—John Burroughs,
  St. Nicholas, 1891

Contents

xiii   Preface
xv   Acknowledgments
xvii   Please Do Not Dig Native Orchids

Part One. Wild Orchids and the Southern Appalachians

3   Introduction
7   What Is an Orchid?
13   Rarity among Orchids
17   Looking for Wild Orchids
30   Special Orchid Places in the Southern Appalachians
37   Preserving Wild Orchids and Their Habitat
43   Using This Book

Part Two. Plant List

53   List of Native Orchid Genera of the Southern Appalachians
55   *Aplectrum hyemale*, puttyroot orchid

Arethusa bulbosa, arethusa or bog rose	57
Calopogon, the grass-pink orchids	60
C. tuberosus, grass-pink orchid	61
Cleistes, the rosebud orchids	64
C. bifaria, smaller rosebud orchid	64
Coeloglossum viride var. *virescens*, green frog orchid	68
Corallorhiza, the coralroot orchids	70
C. bentleyi, Bentley's coralroot orchid	71
C. maculata, spotted coralroot orchid	76
C. odontorhiza, autumn coralroot orchid	80
C. trifida var. *verna*, northern coralroot orchid	83
C. wisteriana, Wister's coralroot orchid	86
Cypripedium, the lady's slippers	90
C. acaule, pink lady's slipper	91
C. candidum, small white lady's slipper	96
C. kentuckiense, Kentucky lady's slipper	99
C. parviflorum, yellow lady's slipper	103
C. reginae, showy lady's slipper	107

111	*Epipactis*, the helleborine orchids
111	*E. helleborine*, helleborine orchid
113	*Galearis spectabilis*, showy orchis
117	*Goodyera*, the rattlesnake plantains
118	*G. pubescens*, downy rattlesnake plantain
120	*G. repens* var. *ophioides*, lesser rattlesnake plantain
123	*Hexalectris*, the crested coralroot orchids
123	*H. spicata*, crested coralroot orchid
126	*Isotria*, the whorled pogonias
126	*I. medeoloides*, small whorled pogonia
130	*I. verticillata*, large whorled pogonia
132	*Liparis*, the twayblade orchids
133	*L. liliifolia*, lily-leaved twayblade
136	*L. loeselii*, Loesel's twayblade
138	*L.* ×*jonesii*, Jones's twayblade
140	*Listera*, the true twayblade orchids
140	*L. cordata*, heart-leaved twayblade
144	*L. smallii*, Small's twayblade
147	*Malaxis*, the adder's mouth orchids
148	*M. unifolia*, green adder's mouth orchid
150	*Platanthera*, the fringed orchids
153	*P. blephariglottis* var. *conspicua*, white fringed orchid
155	*P. ciliaris*, yellow fringed orchid
157	*P. clavellata*, club-spur orchid
159	*P. cristata*, crested fringed orchid
160	*P. flava*, tubercled orchid
162	*P. flava* var. *flava*, southern tubercled orchid
162	*P. flava* var. *herbiola*, northern tubercled orchid
164	*P. grandiflora*, large purple fringed orchid
168	*P. integra*, yellow fringeless orchid
170	*P. integrilabia*, white fringeless orchid
172	*P. lacera*, ragged fringed orchid
175	*P.* ×*andrewsii*, Andrews's fringed orchid
175	*P.* ×*keenanii*, Keenan's fringed orchid
178	*P. leucophaea*, eastern prairie fringed orchid
181	*P. orbiculata*, pad-leaf orchid
184	*P. peramoena*, purple fringeless orchid
189	*P. psycodes*, small purple fringed orchid

Pogonia ophioglossoides, rose pogonia	192
Ponthieva racemosa, shadow witch orchid	195
Spiranthes, the ladies' tresses	197
S. cernua, nodding ladies' tresses	198
S. lacera, slender ladies' tresses	201
S. lacera var. *gracilis*, southern slender ladies' tresses	202
S. lacera var. *lacera*, northern slender ladies' tresses	202
S. lucida, shining ladies' tresses	203
S. magnicamporum, Great Plains ladies' tresses	206
S. ochroleuca, yellow ladies' tresses	208
S. ovalis var. *erostellata*, oval ladies' tresses	210
S. tuberosa, little ladies' tresses	212
S. vernalis, spring ladies' tresses	214
Tipularia discolor, crane-fly orchid	216
Triphora trianthophora, three-birds orchid	219
Glossary	223
Bibliography	227
Index	231

Preface

Although my formal educational background is in history and English, both of which I formerly taught on a secondary level, I have been a naturalist all my life. With summers free, I have had time to pursue my many interests, and I have pursued them ravenously. Even as a young boy growing up in the small town of Pulaski in southwest Virginia, I started making lists of everything I saw: birds, trees, insects, rocks. My first experiences in nature outside my neighborhood came on family hunting trips. I was a grown man before I realized that the real fascination for me was with just being in the woods—the hunting, not the killing. It was a pivotal day indeed when I traded my shotgun for a camera. With such beauty all about me in the forest, I naturally turned to photography in order to record that beauty.

I have had no regrets. And the number of my visits to the woods (to say nothing of my enjoyment of them) has by no means diminished—in fact, quite the contrary. I enjoy much more freedom in the forest with no permits to buy and no seasonal limitations or restrictions on the number of pictures I can take. I feel safer, too. Although the camera does "fire" unintentionally sometimes, it never harms anything when it does.

Although my fascination with the outdoors began when I was a boy, it was later, while giving lectures and leading walks as park naturalist at Claytor Lake State Park, that I came to realize the significance that nature has in my life. For 30 years now I have seriously and studiously enhanced my relationship with nature while continuing to lead hikes and speak to a variety of organizations: bird clubs, wildflower societies, garden clubs, and civic groups. The majority of my free time has been spent studying plants, and I have discovered many county records within a number of states, several state records, and even one world record. But I also have discovered myself. Wandering the woods and fields is something I could do every day of my life were it not for life's ordinary demands. When I think of my quest for native orchids over the last several decades, I would change only one thing: I wish I could have done more of the same.

Since 1975, I have turned the majority of my attention to the study of North American native orchids. The trail of the native orchid has taken me from the barren limestone shores of Newfoundland to the emerald green slopes of Kodiak Island in Alaska, from the sweltering savannahs of the Green Swamp of North Carolina's coast to the redwood forests of northern California, and to hundreds of places in between. There are now more than 140 native orchid species and varieties listed in my files. But, more important, they are also indelibly etched in my mind, each one a unique story, a separate recollection of adventure, a treasured memory. Although I have crossed the continent in search of native orchids, my strongest sentiments spring from my search for wild orchids in the mountains where I grew up, the mountains I have known intimately all my life, the southern Appalachians. This book is an accounting of the native orchids that I have come upon in the southern Appalachian Mountains.

Acknowledgments

No author could ever complete a book of this type without the assistance of many other individuals. The kindness and unselfishness of others have been essential to my having seen and photographed all of the orchids of the southern Appalachians. Their suggestions have made this book more complete and the writing of it much more manageable. In most cases, these individuals have accompanied me in the field and their companionship has greatly enriched my life in many ways. I shall always be grateful for the knowledge gained and even more thankful for the many good friends acquired.

Several persons have been of tremendous help in offering manuscript suggestions. For their time, encouragement, and valuable comments, I extend sincere thanks to Dr. John V. Freudenstein of Ohio State University, Columbus, Ohio; Dr. Hal Horwitz of Richmond, Virginia; Phil Keenan of Dover, New Hampshire; J. Dan Pittillo of Western Carolina University, Cullowhee, North Carolina; and Tom Wieboldt, assistant curator of the Massey Herbarium at Virginia Tech, Blacksburg, Virginia. Special thanks too to Dr. Charles J. (Chuck) Sheviak of the

State University of New York at Albany for kindly taking the time to answer my many questions over the years. For the generous use of the Massey Herbarium at Virginia Tech, thanks also to Dr. Duncan Porter, curator.

The aid I've received in finding orchids all across the southern mountains has been a paradigm of sharing. So many people along the way have shared their orchid sites, their friendship, and their good humor that writing their names on a sheet of paper seems such a small way to thank them. I hope that these friends realize how much they have meant to me and that the listing of their names is but a token of my affection for them. I must first acknowledge three deceased friends—J. T. Baker of Huntsville, Tennessee; the Reverend Joe Hardy of Blacksburg, Virginia; and J. I. (Bus) Jones of Chattanooga, Tennessee—all of whom left to me a legacy of goodwill and a keener appreciation for our beautiful world. For site information, I thank Christina Bird-Holenda of Nipomo, California; Cory and Shirley Curtis of Rollinsford, New Hampshire; Sam and Dora Lee Ellington of Roanoke, Virginia; Bill Grafton of West Virginia University, Morgantown, West Virginia; Dennis Horn of Tullahoma, Tennessee; Dr. Hal and Helen Horwitz of Richmond, Virginia; Fred Huber of the George Washington and Jefferson National Forests, Roanoke, Virginia; Phil Keenan of Dover, New Hampshire; Ed and Lana Mills of New York City; Dr. Larry Morrison of O'Fallon, Illinois; Doug Ogle of Virginia Highlands Community College, Abingdon, Virginia; Professor John (Jack) Payne of Lexington, Kentucky; John Roth of Cave Junction, Oregon; Al Shriver, Scott Shriver, and Clete Smith, all of Pittsburgh, Pennsylvania; Rob Sutter of Raleigh, North Carolina; George R. (Bobby) and Frieda Toler of Roanoke, Virginia; and Charles (Chuck) Wilson of Hixson, Tennessee. And thanks also to my good friend Eddie Goodson, who has tolerated many times my turning our trips to explore old, abandoned railroads into orchid hunting trips.

One of the dear friends acknowledged above died while this book was in press. For Bobby Toler, who was a faithful orchid hunting companion and best friend for more than a decade, I offer a special word of remembrance.

Please Do Not Dig Native Orchids

At some point, almost everyone who is interested in plants gets an urge to garden. Most of these people are satisfied with ornamental plants, but many get their enjoyment from a wildflower garden. For them, the seemingly natural progression is to transplant wildflowers to the home garden. Many wildflower guides even give instructions on wildflower gardening. However, education as to just which plants should be transplanted to such a garden is the most important bit of knowledge a would-be gardener can acquire. There are a number of sound reasons why the disturbance of our wild orchids is not an acceptable practice.

The primary reason is simply one of preservation. The digging and removal of an orchid plant from its natural habitat will almost surely result in the demise of the plant. Each species of orchid has its own special relationship with a fungus partner that is responsible for the survival of that orchid. Unless that particular fungus partner can be replicated in the home garden, no orchid will survive after the depletion of the incidental supply of fungus brought along with the transplantation. Furthermore, in many cases, orchid roots are attached to the fungus partner by fragile, thin filaments, which are invariably damaged in any digging efforts.

Second, on the basis of general moral and ethical principles it is important that wild orchids not be disturbed. Orchids are intricately designed and particularly beautiful plants. In addition, many orchid species are exceptionally rare. Removing these plants or disturbing their natural habitat is a singularly selfish act. Leave the plants so they can survive and reproduce naturally. Leave them for others to enjoy. Leave the plants just because they are there. One must make the moral decision to allow nature to dictate the future of such beautiful and rare plants.

Finally, in many areas, the native flora is protected by law. A considerable number of native orchids are on either federal or state endangered or threatened species lists. Such listing makes any disturbance of these plants illegal. Many plants, orchids included, are located in parks or national forests whose very purpose is to protect the flora from disturbance of any kind.

Educate yourself by learning to identify and differentiate the species of native orchids and the life processes that make them special plants. Make a decision to become an advocate of protection, become active in the preservation of our natural heritage, and forgo the urge to transplant any wild orchid. You will be better for it and so will the orchids.

Part One Wild Orchids and the Southern Appalachians

Introduction

Whether on a corsage at some formal occasion or in a display at the local shopping mall, the beauty of orchids is acknowledged by everyone. The enchanting flowers we call orchids are known throughout the world. For most people, however, their name conjures up a vision of exotic rain forests or, at least, steamy greenhouses. Few people realize that there is a marvelous population of wild orchids found outside tropical regions, and even fewer people know that dozens of species grow in the southern Appalachian Mountains.

Wild orchids exist worldwide, from the equator to the arctic tundra. In the mid-eighteenth century, when the botanist Carolus Linnaeus was redefining the way botanists would name plants, there were only about 65 species of orchids known to science. By the mid-nineteenth century, the number of named species of native orchids had grown to over 5,000. Today there are estimated to be between 20,000 and 25,000 orchid species worldwide, many of which are yet to be named. And the number is growing. Botanists are still discovering new species, particularly in the widely biodiverse tropic regions. On the North American continent, two new species of orchids have been discovered since 1988 in the Canadian province of Newfoundland, and this book describes a new species found by the author in the state of West Virginia in 1996. Altogether, there are in North America, excluding Mexico and the subtropical parts of Florida, approximately 150 species and varieties of native orchids.

This book deals with only a portion of North America, of course—namely, western Virginia and western North Carolina and the eastern regions of West Virginia, Kentucky, and Tennessee. The crests of the Blue Ridge range define the eastern boundary of the southern Appalachians and become the eastern limit of the range of this book. The ridges of the Allegheny Mountains define the western boundary. The northern boundary of the area covered here can generally be considered as the line drawn by the Potomac River as it descends from northeastern West Virginia, separating Virginia from Maryland. The southern boundary is the line that separates northern Alabama and northern Georgia from southern Tennessee and southwestern North Carolina, approximately 35° north latitude. Within this general area, the county-by-county distribution of each species of orchid discussed in the book is specified on a range map.

Within the boundaries of the area treated in this book is the land between the Blue Ridge and the Alleghenies, which is called the Ridge and Valley Province. In Virginia, this province encompasses the Shenandoah Valley and parts of the James, Roanoke, and New River Valleys of southwest Virginia. This same physiographic region in Tennessee follows the margins of the Tennessee River drainage southwestward between the Great Smoky Mountains to the east and the escarpment of the Cumberland Plateau to the west.

The Cumberland Plateau is a southwestern extension of the Alleghenies and includes parts of eastern Tennessee, southwestern West Virginia, and southeastern Kentucky. In addition to a large number of orchid species typically found in the mountains, several wonderful and special orchids that have affinities with orchids more commonly located in the coastal plain are at home on the Cumberland Plateau.

According to geologists, the Appalachian chain of mountains once extended from the rugged hills of Ireland to the low hills of Texas. After the separation of the continents that began some 200 million years ago and after tens of millions of years of mountain construction and mountain erosion through a parade of changing climates, the Appalachians are presently more clearly represented as existing in two parts. The northern Appalachians begin on the island of Newfoundland and reach southward through New England to the coalfields of eastern Pennsylvania. At this point, the Appalachians begin to split into two distinct ranges that define the southern Appalachians.

The Blue Ridge range extends basically in a north-northeast to south-southwest direction from the Virginia-Maryland border to the Virginia–North Carolina line. In southern Virginia, the Blue Ridge begins to take a decidedly western turn, creating a clearly defined escarpment extending basically along the Virginia–North Carolina border. A bit farther to the west in Virginia lie the promontories of White Top Mountain and Mount Rogers, Virginia's highest peak. As they extend southwest along the western region of North Carolina, the vast ridgelines of the Blue Ridge form the highest mountain range in the eastern United States, including the highest mountain

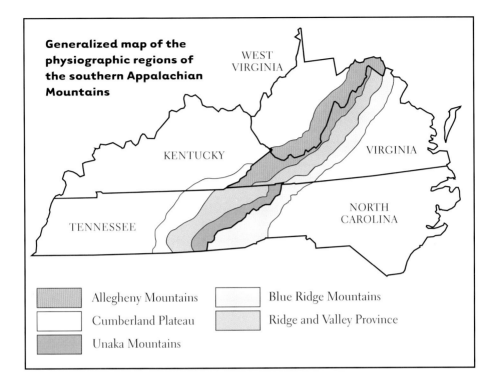

Generalized map of the physiographic regions of the southern Appalachian Mountains

- Allegheny Mountains
- Cumberland Plateau
- Unaka Mountains
- Blue Ridge Mountains
- Ridge and Valley Province

east of the Mississippi River, Mount Mitchell (6,684 feet) located in Yancey County, North Carolina.

The Allegheny mountain range is the western fork of the Appalachians south of Pennsylvania and, like the Blue Ridge, extends in a more or less northeast to southwest direction. Its ridges rise to considerable elevations just west of the border between Virginia and West Virginia, especially along the western edge of the Greenbrier River Valley. Among these high Allegheny crests like Spruce Knob in West Virginia are a number of distinctly northern habitats that have nurtured some of our more interesting orchid species. In southern West Virginia, far southwest Virginia, and eastern Kentucky, the Alleghenies are famous for their rough terrain. The huge layered masses of shales and sandstones that make up the Alleghenies provide a superabundance of acidic habitat of the kind required by most native orchid species. Particularly in southwest Virginia and continuing into neighboring parts of Kentucky and Tennessee, many of these Allegheny ridges are organized in long parallel rows generally oriented northeast to southwest. Capped with deposits of sandstone, these mountains frequently have nearly level crest lines that go on uninterrupted for tens of miles. They are often referred to as the "long ridges." These mountains of sedimentary rock stand in stark contrast to the more jagged peaks of the Blue Ridge, whose rock was born of volcanoes and tremendous heat and pressure.

Both in Virginia and Tennessee, the Ridge and Valley Province provides widely diverse habitats: ridges composed of acidic sandstones and shales and valleys largely made up of vast deposits of limestone. The more basic soils found in these valleys provide ample habitat for native orchids that do not require an acidic environment.

The number of native orchid species found in the southern mountains is greatly enhanced by the legacy of the last Ice Age. Although the actual glacial ice came no farther south than what is now mid-Pennsylvania, it had a tremendous impact on the land to the south. As the ice advanced, plants and animals alike migrated south just ahead of the ice sheet in order to survive. Plants that were native to the far North found themselves existing much farther south. The area covered in this book had a subarctic climate during the Ice Age, and the northern plants were right at home. As the ice finally retreated about 10,000 years ago, North America warmed and, for the most part, the northern plants that had been driven far to the south also migrated back toward the cooler climate of the north. Some northern plants, however, found parts of the southern Appalachians to their liking. The high ridges of the Great Smoky Mountains, which are bisected by the border between North Carolina and Tennessee, the Black Mountains of North Carolina, and the Unakas, also along the Tennessee–North Carolina border, have many regions of spruce and fir trees above an elevation of 5,000 feet. Several peaks reach above 6,000 feet. Many areas of the Monongahela National Forest in West Virginia—like Spruce Knob, Canaan Valley, and Dolly Sods—are well over 4,000 feet in elevation and have typical Canadian zones of spruce forests. Likewise, areas of the Shenandoah National Park in Virginia have spruce habitats. Most of these high-elevation areas have become modern havens for northern plant species, including orchids, that came south ahead of the ice.

In addition, many of these high mountain areas of the southern Appalachians harbor cold, poorly drained valleys where conditions were just right to create distinctly northern habitats called bogs. Some of the bogs have persisted for thousands of years. The beauty of these southern bogs is that many of them shelter some of the more showy species of native orchids that are ordinarily found only in the swampy coastal plain or far to the north in the well-known muskeg bogs.

The southern Appalachian Mountains are a land of immense variety. There are highlands and lowlands; broad, well-watered valleys; steep, narrow valleys beneath dry pine-oak woodlands, where shale and sandstone have created a parched, acidic environment; mighty rivers and tiny rivulets that bring life to the abundant vegetation and carry vegetative remains to the sea. Each region has contributed to the diversity of life within these mountains. This is a land of mature trees and open grasslands, a land of large mammals and small birds, a land of sunshine and rain, a land of spring wildflowers and autumn colors; it is a land of life in myriad forms. There is no other place quite like the southern Appalachian Mountains.

What Is an Orchid?

Officially, orchids are among the class of plants called monocotyledons or monocots for short. Monocots are so called because germinating plants have only one seed leaf or cotyledon. The development of the cotyledon is seldom seen in nature because the process ordinarily takes place underground. Monocot leaves usually have parallel venation. This is in opposition to the dicotyledons, which have two seed leaves and more often display reticulate leaf venation. Monocots normally have flower parts in threes or multiples of three. In the orchids there are three sepals and three petals. On the other hand, dicots usually have flower parts in fours or fives. Monocots include the lily family, Liliaceae. Orchids and lilies are generally believed to be descended from a common ancestor. Among the better-known families also included in the monocots are the iris family, Iridaceae; the grasses, Poaceae; and the sedges, Cyperaceae.

Although orchid flowers have three petals, in the majority of the orchids one of the petals (usually the lowermost) looks nothing like the others. The wondrous and

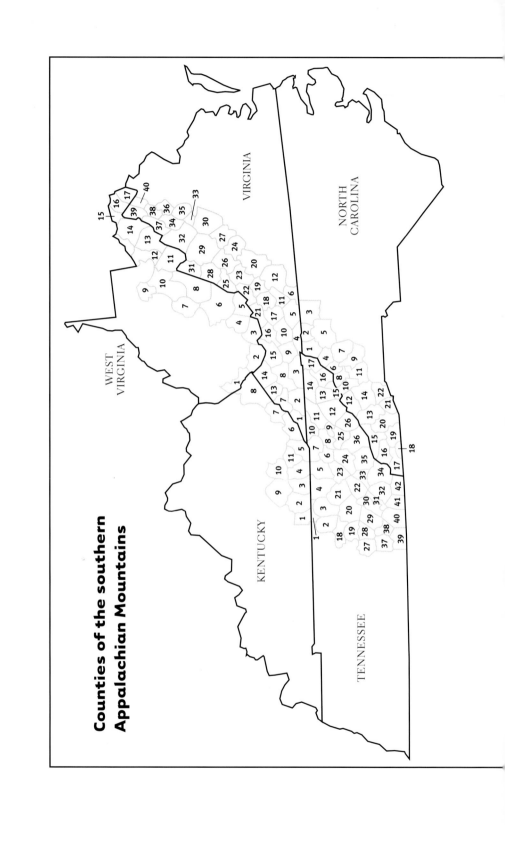

Key

Virginia

1. Lee
2. Scott
3. Washington
4. Grayson
5. Carroll
6. Patrick
7. Wise
8. Russell
9. Smyth
10. Wythe
11. Floyd
12. Franklin
13. Dickinson
14. Buchanan
15. Tazewell
16. Bland
17. Pulaski
18. Montgomery
19. Roanoke
20. Bedford
21. Giles
22. Craig
23. Botetourt
24. Amherst
25. Alleghany
26. Rockbridge
27. Nelson
28. Bath
29. Augusta
30. Albemarle
31. Highland
32. Rockingham
33. Greene
34. Page
35. Madison
36. Rappahannock
37. Shenandoah
38. Warren
39. Frederick
40. Clarke

West Virginia

1. Mingo
2. McDowell
3. Mercer
4. Summers
5. Monroe
6. Greenbrier
7. Webster
8. Pocahontas
9. Barbour
10. Randolph
11. Pendleton
12. Grant
13. Hardy
14. Hampshire
15. Morgan
16. Berkeley
17. Jefferson

Kentucky

1. Clinton
2. Wayne
3. McCreary
4. Whitley
5. Bell
6. Harlan
7. Letcher
8. Pike
9. Pulaski
10. Laurel
11. Knox

Tennessee

1. Pickett
2. Overton
3. Fentress
4. Scott
5. Campbell
6. Union
7. Claiborne
8. Grainger
9. Hamblen
10. Hancock
11. Hawkins
12. Greene
13. Washington
14. Sullivan
15. Unicoi
16. Carter
17. Johnson
18. Putnam
19. White
20. Cumberland
21. Morgan
22. Roane
23. Anderson
24. Knox
25. Jefferson
26. Cocke
27. Warren
28. Van Buren
29. Bledsoe
30. Rhea
31. Meigs
32. McMinn
33. Loudon
34. Monroe
35. Blount
36. Sevier
37. Grundy
38. Sequatchie
39. Marion
40. Hamilton
41. Bradley
42. Polk

North Carolina

1. Ashe
2. Alleghany
3. Surry
4. Watauga
5. Wilkes
6. Avery
7. Caldwell
8. Mitchell
9. Burke
10. Yancey
11. McDowell
12. Madison
13. Haywood
14. Buncombe
15. Swain
16. Graham
17. Cherokee
18. Clay
19. Macon
20. Jackson
21. Transylvania
22. Henderson

seemingly endless variety exhibited by this petal makes the orchid flower so special. It is this special third petal that can assume the golden yellow, pouchlike shape of a yellow lady's slipper or the brilliantly purple, multilacerated shape of the large purple fringed orchid flower. This highly evolved third petal is commonly called the lip or labellum. The wide diversity of orchid petals is directly related to the type of pollinators that reach each species and the method of pollination that they bring to the plant.

Orchid flowers usually go through a process known as resupination. In the bud stage, the lip is uppermost; but as the flower opens, it begins to twist, so that by the time of full flowering, the lip has become lowermost. In some species, such as the pad-leaf orchid, this twisting can easily be seen in the flower stalk or pedicel. Some orchid species do not undergo this twisting process, however, and are called nonresupinate. The best example of a nonresupinate orchid flower in the southern mountains is the grass-pink orchid. Some few species, such as the club-spur orchid, may undergo an incomplete turning process, as a result of which, when the plant blooms, the lips of the various flowers on a single stem are often pointing in several directions.

One of the primary characters by which orchid flowers are separated from other plants is the process by which they present their pollen to pollinators. Most flowers have a centrally extruded "pollen catcher" called a pistil. This is the female part of the flower. The pistil consists of the ovary, the style, and the stigma. The ovary contains the ovules, which, when fertilized, will become the seeds. The ovary develops below the style, which supports the stigma that forms at the tip. The stigma is the part that receives the pollen. The style is usually surrounded by varying numbers of stamens, the male part of the flower. A stamen consists of a threadlike filament at the end of which is the anther. The anther is the pollen-bearing part of the flower. When the plant is ready for fertilization, the pollinator (whether an insect, a bird, or even the wind) transports the ripened pollen from the anther to the pistil, either on the same flower or from the anther of one flower to the pistil of another. Pollen grains germinate on the stigma, and the resulting sex cells move through the style to reach the ovules, where fertilization takes place and seed formation begins.

In the orchid flower, the pistil and two stamens are fused together into a single unit referred to as the column. The column is located opposite the lip. Pollen is contained in a tiny mass called a pollinium, which is attached by means of a stalk or stipe to a sticky gland or pad, together making up a pollinarium. The pollinaria are presented on either side of the column. When insects like bees, beetles, wasps, ants, flies, and moths rub against the gland, the pollinarium will adhere to the insect and be carried to another flower on the same plant or a flower of another orchid plant of the same species. Thus the pollinator is able to transport a considerable quantity of pollen in a single visit. The more pollen delivered, the more chance of fertilization. In the case of orchids, which are known to produce tremendous numbers of ovules within the ovary, this transportation of a large quantity of pollen grains makes the entire process more efficient. In some species, the position of the pollinaria is on either side of an opening to the nectary, luring the pollinator in so that it must press against the gland.

In some species, such as the fringed orchids, the pollinaria are cleverly designed so that even a near-miss can still result in successful fertilization. In this instance, the pollinaria are extruded, and a pollinator need barely touch the sticky gland in order for them to adhere. But that's not all. The pollinaria in the fringed orchids (among others) are shaped in an elongated, teardrop fashion, with the smaller end forward and down. This smaller end is a gland called a viscidium. The viscidium has a tremendously efficient adhesive substance that immediately attaches fast to the insect. The pollinarium is then pulled away easily (sometimes so easily, in fact, that even before it can be picked up by a pollinator it adheres to the lip of the flower as the flower bud unfolds). The mature pollinarium attached to a pollinator is then ready to be transported. In looking at the reproductive structure of orchids, I have often inadvertently touched a pollinarium. The small end will adhere tightly to my skin. When I try to dislodge it, the large end (the pollen) will crumble like powder, while the adhesive end continues its tenacious hold. Through this marvelous system orchids ensure their own propagation.

In many orchid species, when the flower is pollinated, it begins to wither or show some sign such as a drooping petal or lip to advertise the fact that it is no longer open

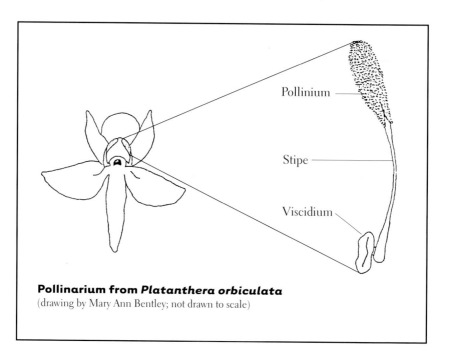

Pollinarium from *Platanthera orbiculata*
(drawing by Mary Ann Bentley; not drawn to scale)

for business. In some orchids, the actual fertilization may not occur for some time after pollination. Still other species, like the true twayblades, may hold onto their petals long into the development of the fertilized ovary. The swollen ovary located behind the flower becomes the seed-bearing capsule, which, when ripe, will burst and scatter the seeds. Generally there are thousands of seeds in one orchid capsule. Some tropical species may have over a million seeds. The production of huge quantities of seeds is an orchid's way of helping to ensure the reproduction of its kind. With so many seeds out there, some are sure to find a favorable spot to germinate and produce more orchids. It is a game of numbers. The more seeds, the more chances there are that some of them will end up in a suitable habitat.

Orchid plants are dependent upon a relationship (mycorrhiza) with a fungus partner. The tiny seeds produced by orchids do not carry much life-sustaining material, so the fungal relationship is very important in helping to supply the young orchid plant with the nutrients it needs for survival and in sustaining the mature plant. The difficulty of maintaining this relationship is the primary reason why attempts to transplant native orchids are so often

A multitude of tiny seeds, typical of plants in the orchid family, from an opened capsule of a spotted coralroot orchid flower.

unsuccessful. In the genus *Corallorhiza*, the coralroot orchids, this alliance between the rhizomes and fungi is most pronounced. In these orchids, which have no green leaves and thus very little chlorophyll for photosynthesis, the fungus is of utmost importance, as it helps convert energy from outside sources for use by the plant. It is a sure bet that without its fungus partner an orchid will not prosper.

Rarity among Orchids

Rarity among orchids can be a relative matter. I once had a friend tell me, "If you can't find it, then it's rare enough." In a simplistic way, my friend could not have been more correct. When I first became fascinated with native orchids, I searched through all of three seasons for yellow lady's slipper and found only two plants in the vicinity of my home. I had reached the conclusion that yellow lady's slipper was indeed rare. But after a hiking trip on the Appalachian Trail in neighboring Bland County, where I counted 242 yellow lady's slipper plants,

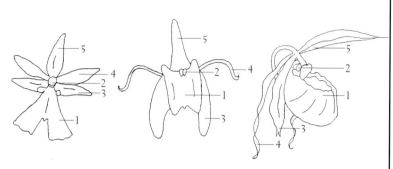

Listera smallii, Small's twayblade orchid. Note the knobs at the base of the lip.

Malaxis unifolia, green adder's mouth orchid, the smallest orchid flower in the southern Appalachians.

Cypripedium kentuckiense, Kentucky lady's slipper, showing the joined lateral sepals (synsepal).

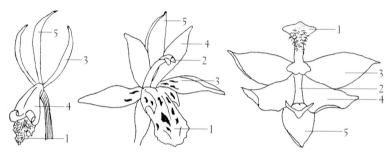

Cleistes bifaria, smaller rosebud orchid. The column is inside the tubelike structure of the petals.

Corallorhiza maculata, spotted coralroot orchid. Note the small, toothlike lobes of the lip.

Calopogon tuberosus, grass-pink orchid. The lip is uppermost (thus the flower is called nonresupinate).

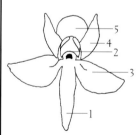

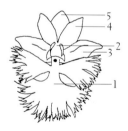

Platanthera orbiculata, pad-leaf orchid. Note the unlobed lip.

Platanthera grandiflora, large purple fringed orchid. Note the fringed, three-lobed lip.

Corresponding floral parts in various orchid species of the southern Appalachians

(drawings by Mary Ann Bentley; not drawn to scale)

Key: 1. Lip
2. Column
3. Lateral sepals
4. Lateral petals
5. Dorsal sepal

my opinion changed. In later years, as I traveled on the Bruce Peninsula in Ontario and the Great Northern Peninsula of Newfoundland, where I encountered literally thousands of yellow lady's slipper plants, my thoughts changed yet again. It became obvious that whether or not yellow lady's slipper was rare depended upon where I was.

Some native orchid species that grow in the southern Appalachian Mountains are particularly common. Downy rattlesnake plantain is found in almost every patch of woods in almost every type of habitat throughout the area. Much the same is true of pink lady's slipper. There probably isn't one spring turkey hunter or trout fisherman in the East who has not passed in close proximity to plants of pink lady's slipper, though the plants may have gone unnoticed or at least unrecognized as orchids. Of the 52 orchid species described in this book, 12 (23 percent) exist in more than 75 of the 132 counties of the southern Appalachians. That means that nearly one-fourth of the species should be rather easily located.

But make no mistake: there are rare orchid species in the southern mountains, several of them. A full 25 percent, or 13 of the 52 species, are found in fewer than seven counties. Six of the species are found in only one county.

Rarity can also be a matter of timing. Take my pursuit of the small whorled pogonia, for instance. I tried several avenues to gain access to this orchid, including contacting one of the authors of a regional floral guide. I found this author to be quite open and cooperative in our first few letters, but when I asked about the site for the small whorled pogonia that had been listed in his book, I never heard from him again. My luck seemed to be getting better when I heard that several new, well-populated sites had been found in New England. To take advantage of this, I arranged to meet Phil Keenan of New Hampshire on his home turf to look for the orchid in 1992. But as it turned out, we were able to locate only one plant in flower in New Hampshire and one plant in Maine, where there had been a hundred plants in bloom the previous year. So rarity can be relative to being in the right place at the right time, not only the right time of season but also in the right season. People lucky enough to have found the heavily populated sites that were discovered in New England in the 1980s would have a different perspective on the rarity of the small whorled pogonia from mine.

Orchids do not necessarily meet the criteria by which

we might evaluate other plant families. Many native orchids have special quirks all their own, the understanding of which can help explain their rarity and even help illuminate why some species may seem scarce when they really are not. For instance, it takes a new lady's slipper plant some three or four seasons before it stores enough energy to send up a flowering stem. The rattlesnake plantains initially produce a rosette of leaves, but they may not send up a flowering stem for five or six years. The three-birds orchid is particularly sensitive to changes in the weather: it is believed that the plants will not open their flowers until some 48 hours after the passage of a cold front in early August. Along with several other species, three-birds orchids may lie dormant underground through several seasons before storing enough energy to appear above the ground.

Some orchids, because of their lack of bright colors or their well-camouflaged shape, go virtually unnoticed. The large whorled pogonia is actually rather common in the spring but is often overlooked because its narrow, green and yellowish petals give it a natural camouflage. Loesel's twayblade, commonly looked on as a northern orchid, is much more prolific in the southern Appalachians than once thought. The problem in finding this orchid is that, even when in full bloom, it is a very tiny plant whose miniature green flowers are well hidden among grasses of exactly the same color.

Rarity can also be a matter of opportunity. By the time late summer arrives, all but the most tenacious botanists have moved on to activities other than wildflower searches. The weather is hot and there are ticks and mosquitoes with which to contend, so fewer walks are taken in the woods and fields. Yet it is at this time of year that a number of orchids flower. All but one of the eight species of ladies' tresses are late-summer- or fall-blossoming plants. Some species of ladies' tresses are quite common, yet I have heard many longtime wildflower enthusiasts say they have never seen one. Is it the plants that are rare or the excursions to find them?

Knowledge of the presence of wild orchids in the southern Appalachian Mountains is not widespread. The picture of orchids in most people's minds is corrupted by their association with greenhouses and large snakes hanging from vines in the jungles of Central America—where orchids "should grow." The consensus of opinion is undoubtedly, "Well, if wild orchids are found around these parts, they must be pretty rare. I've never seen one." I have had many outdoor people say to me, "Oh yeah, I know what a lady's slipper is, but I didn't know that it's an orchid." If one looks at the right time and in the right place and has the perseverance to find them, there are dozens of species of wild orchids to be discovered in the southern mountains. The general perception that wild orchids are rare is not a particularly valid one.

Looking for Wild Orchids

When should one look for orchids in the southern Appalachian Mountains? Believe it or not, the answer is anytime. When the heavy frosts and hard freezes of November and December begin to descend over the ridges and valleys, hunting wild orchids is probably the farthest thing from the minds of most people. But it is about this time that at least two species of orchids become easier to find than at any other time of the year, including bloom time. By mid-autumn, both the puttyroot orchid and the cranefly orchid begin to grow green, showy leaves that will stand out against the backdrop of the newly brown, almost colorless landscape of the winter woods. Both of these orchid species are inconspicuously colored when they bloom, making them difficult to locate in their habitat under a dark canopy of trees. But the winter leaves are easier to see and add a bright note to a brisk winter walk. One can simply mark the location of the leaves in the winter and save oneself a great deal of effort locating the plants when they do bloom in the warmer weather.

When to Look

Dried seed capsules or pods seen in the winter are often "dead" giveaways to certain species. Many of these skeletonlike pods survive into the next blooming season. Loesel's twayblade is so tiny and so well camouflaged when it blooms in early summer that it can be exasperating to locate, even when one thinks he knows where the plants are. But in the early fall, when the grasses begin to die away, the seed pod of Loesel's twayblade takes on a bright greenish-yellow color that easily stands out against the usually drab backdrop of its wet ditch habitat. Capsules from the lady's slippers, showy orchis, puttyroot, and other orchids are also distinctive. I often use these discoveries of winter capsules to help me know where I should look for orchids in the next flowering season.

Of course, the warm season is still the best time to find most species of wild orchids in the southern mountains. Flowering wild orchids can be found in the area from late March into early November. Wister's coralroot orchid is the earliest orchid to flower as spring first comes to the southern reaches of Tennessee and North Carolina. The enchanting showy orchis is next as spring begins to creep both northward and up the ridges. I have made many early spring excursions to the Great Smoky Mountains to look for showy orchis in Greenbriar Cove and along the Roaring Fork Motor Nature Trail. This has always been a welcome ritual for breaking the hold of the cold days past. The lady's slippers are the next bloomers in the woods, dominating the thoughts of the orchidophile through the month of May. By early summer, the leafless stems of the coralroot orchids appear. At least one of the five coralroot orchid species is in flower in the southern mountains during each part of the growing season, from March into October.

With the closing of the forest canopy and the arrival of summer weather, the blossoming of most of the orchid species of the southern Appalachians moves from the woods to the meadows and wet ditches. The fringed orchids, with their tall, brightly colored racemes of flowers, decorate meadows from mid-June into early August. The several species of ladies' tresses embellish roadsides, wet fields, and the multicolored edges of woods in the autumn. Nodding ladies' tresses can even be found in the heavy frosts of early November in the higher regions of the southern Appalachian Mountains.

Of course, as with any plants, the blooming time for orchids in the southern mountains can be influenced by weather extremes such as cold springs or summer droughts. On October 3, 1999, after three consecutive very dry summers, I found 56 plants of spotted coralroot orchid in prime bloom in Giles County, Virginia. Some plants at this site still had fresh flowers into November. What makes this a particularly strange occurrence is that the spotted coralroot orchid normally blooms in late July.

Orchid hunting in the southern Appalachians can provide year-round enjoyment. With Scott and Al Shriver and Clete Smith, I once waded Anthony Creek in West Virginia to see a location for lesser rattlesnake plantain;

we readily found the plant's leaf rosettes after braving the cold, cold, waist-deep water in hip boots—all on the last day of November. For those of us who don't know any better, neither is orchid hunting restricted to daylight. Following my friend Chuck Wilson closely, so as not to get lost, I have actually hunted orchids with a flashlight well after dark in, of all places, a cemetery on the Chickamauga battlefield in northern Georgia. Whether one is casting off the cold from a boring winter day, warming to the new spring sun, strolling through summer meadows, or catching the last clean, crisp days of autumn, a commune with nature can include a pleasurable liaison with wild orchids in our southern mountains. Pack up a field guide, fortify yourself with enthusiasm, and be ready to venture into an experience that can be rewarding throughout a lifetime.

Where to Look

Native orchids can be found just about anywhere in the southern Appalachian Mountains. Several years ago, while mowing my backyard, I came upon two plants of slender ladies' tresses. I had no idea how they got there, but I relished the fact that I had orchids growing in my own yard and that I had not transplanted them there.

Native orchids of one species or another can be found in almost any woodlot. Downy rattlesnake plantain is present in all types of forests, from acid evergreen woods to woods mixed with oak and other hardwoods, to woods with basic soils. I have seen pink lady's slipper plants at the top of Spruce Knob, West Virginia's highest elevation at 4,861 feet. And I have seen them along the New River in Virginia at just over 1,800 feet. Pink lady's slipper is one of the finest examples of the beauty of wild orchids and, at the same time, one of the more common species in the southern Appalachians, thus available to anyone who wants to take a woodland stroll in mid-spring.

Most of the orchids in the southern mountains are acid-loving creatures, right at home in the Allegheny ridges composed of acid-producing shales and sandstones. But it is not just the soil itself that dictates a plant's environment. For example, bogs are special habitats composed of decayed organic or vegetative material. Decaying fern and moss species give true bogs, no matter where they are found, an inherently highly acidic environment.

Some orchids in the southern Appalachians neverthe-

Flowering Periods for the Native Orchids of the Southern Appalachian Mountains

Orchid	Flowering Period*						
	Apr.	May	June	July	Aug.	Sept.	Oct.
Aplectrum hyemale Puttyroot orchid		■					
Arethusa bulbosa Arethusa			■				
Calopogon tuberosus Grass-pink orchid				■			
Cleistes bifaria Smaller rosebud orchid			■				
Coeloglossum viride Green frog orchid		■					
Corallorhiza bentleyi Bentley's coralroot orchid					■		
*C. maculata*** Spotted coralroot orchid				■			

Note: Orchids are listed in the same order in which they appear in the text.
*Flowering periods reflect early dates at lower elevations and/or lower latitudes and later dates at higher elevations and/or higher latitudes.
**See text to determine flowering periods for hybrids and forms.

Flowering Period*							Orchid
Apr.	May	June	July	Aug.	Sept.	Oct.	
					▓	▓	*C. odontorhiza*★★ Autumn coralroot orchid
		▓					*C. trifida* Northern coralroot orchid
▓	▓						*C. wisteriana*★★ Wister's coralroot orchid
	▓	▓					*Cypripedium acaule* Pink lady's slipper
	▓						*C. candidum* Small white lady's slipper
	▓	▓					*C. kentuckiense* Kentucky lady's slipper
	▓	▓					*C. parviflorum* Yellow lady's slipper
		▓					*C. reginae* Showy lady's slipper
		▓					*Epipactis helleborine* Helleborine orchid

Orchid	Flowering Period*						
	Apr.	May	June	July	Aug.	Sept.	Oct.
Galearis spectabilis Showy orchis	▨	▨					
Goodyera pubescens Downy rattlesnake plantain				▨			
G. repens Lesser rattlesnake plantain				▨	▨		
Hexalectris spicata Crested coralroot orchid				▨	▨		
Isotria medeoloides Small whorled pogonia		▨					
I. verticillata Large whorled pogonia		▨					
Liparis liliifolia Lily-leaved twayblade			▨				
*L. loeselii*** Loesel's twayblade			▨				
Listera cordata Heart-leaved twayblade		▨					

Flowering Period*							Orchid
Apr.	May	June	July	Aug.	Sept.	Oct.	
							L. smallii Small's twayblade
							Malaxis unifolia Green adder's mouth orchid
							Platanthera blephariglottis White fringed orchid
							P. ciliaris Yellow fringed orchid
							P. clavellata Club-spur orchid
							P. cristata Crested fringed orchid
							P. flava var. *flava* (s) Tubercled orchid *P. flava* var. *herbiola* (n)
							P. grandiflora Large purple fringed orchid
							P. integra Yellow fringeless orchid

Orchid	Flowering Period*						
	Apr.	May	June	July	Aug.	Sept.	Oct.
P. integrilabia White fringeless orchid					■		
P. lacera** Ragged fringed orchid			■				
P. leucophaea Eastern prairie fringed orchid			■				
P. orbiculata Pad-leaf orchid			■	■			
P. peramoena Purple fringeless orchid				■			
P. psycodes Small purple fringed orchid			■	■			
Pogonia ophioglossoides Rose pogonia			■				
Ponthieva racemosa Shadow witch orchid						■	
Spiranthes cernua Nodding ladies' tresses						■	■

Flowering Period*	Orchid
Apr. | May | June | July | Aug. | Sept. | Oct.	
July–Aug.	S. lacera var. gracilis (s) Slender ladies' tresses
July	S. lacera var. lacera (n)
May–June	S. lucida Shining ladies' tresses
Sept.	S. magnicamporum Great Plains ladies' tresses
Aug.–Sept.	S. ochroleuca Yellow ladies' tresses
Aug.–Sept.	S. ovalis Oval ladies' tresses
July–Aug.	S. tuberosa Little ladies' tresses
July–Aug.	S. vernalis Spring ladies' tresses
July–Aug.	Tipularia discolor Crane-fly orchid
Aug.	Triphora trianthophora Three-birds orchid

less prefer limestone or more basic soils. These plants are often called calciphiles, in reference to the high calcium content of the soil in which they grow, and they include Wister's coralroot orchid, showy orchis, and crested coralroot. In the Greenbrier Valley of eastern West Virginia, in many areas of Virginia, including the great Shenandoah Valley, and in the Tennessee River Valley of eastern Tennessee, the southern mountains are blessed with extensive limestone regions in which calciphiles flourish.

Disturbed areas, even waste places, can often be among the finest of orchid habitats. In August 1979, as I was driving through the small community of Cowen, West Virginia, something tall and pink drew my eyes to the left. I pulled over, right in the middle of town, and walked across the street into a vacant lot beside a service station. To my amazement, there were over a dozen blooming plants of purple fringeless orchid. Open fields or meadows, particularly wet ones, are some of the easiest places to find certain orchid species in mid- to late summer. In just one tiny field along the Blue Ridge Parkway in Alleghany County, North Carolina, I have seen up to 155 blooming plants of yellow fringed orchid, several plants of ragged fringed orchid, spring ladies' tresses, and slender ladies' tresses, a number of plants of the tiny green adder's mouth orchid, and, in the spring, even a few pink lady's slipper plants.

Orchid habitats also include roadside ditches. Many species find their favored niche in ditches, but sometimes at their own peril. I know of a location for shining ladies' tresses, rare in the southern mountains, that was discovered in southwest Virginia in 1989. But this site, in Giles County, is in a persistently muddy ditch not two feet from the pavement. Just one errant tire off the road could easily destroy this orchid population.

So native orchids can be found in many habitats: woods, fields, meadows, bogs, and ditches. I cannot think of any type of land that does not harbor some species of wild orchid somewhere in the southern Appalachians. Even in towns, they can sometimes be found in woodlots that people routinely overlook. Just take a walk there sometime; you might be surprised at what you find. Put on a little insect repellent and venture out into a late summer meadow instead of sitting in the blast of that 10,000 BTU air-conditioner. Walk a mountain ridge and step off the path a few times. There are secrets you can't imagine, and there are wild orchids too. And, oh yes, look around your yard before you fire up that belching mowing machine. You never know!

How to Look

My first advice to someone who wants to hunt orchids would be to get out into the field. Go often, and return to the same places in more than one season since, as already discussed, the blooming habits of wild orchids can be very erratic. The number of visible plants at any one location can vary tremendously from year to year. Nothing can take the place of comprehensive fieldwork. Besides, for me at least, being out of doors is the most enjoyable part of learning about wild orchids.

Second, talk with someone who knows wildflowers, particularly someone well versed in the flower of your choice, whether it's orchids or another plant family. Write to people; don't be shy in asking questions. Most people are more than willing to share information, especially if you can offer them useful information in exchange. Earn a good reputation as an ethical botanist. You might be surprised how a good name can follow you and be of tremendous help, particularly in dealing with a new contact. Try corresponding with the naturalist staff at state and national parks. Visit herbaria at colleges and universities and let it be known what data you are seeking. While at the different herbaria, ask to look at their specimens for the plants you have in mind. The collection information on the herbarium sheet can be an invaluable help in locating species in the wild. Be inventive. I once stopped at a library in a town I had never visited and asked who in town was knowledgeable about wildflowers. The librarian led me to a man who showed me one of the more rare species of wildflowers in the eastern United States: a plant called Shortia, locations for which were once lost for over a hundred years. I have friends from Pittsburgh who visited a church service with snake handlers in order to talk with the man who gathered the snakes. My friends figured that, since this man was out in the woods a great deal, he might be able to help locate certain species of wild orchids. Use your imagination.

Next, it is very important to acquire and keep good maps. There are a number of sources for maps, including the United States Geological Survey, state and local highway departments, state and federal parks or recreational areas, and outdoor stores. With the explosion of interest in hiking, biking, canoeing and the like, most "outfitter" stores now carry a good stock of useful maps as well as other supplies. Many good books of maps are available nowadays. The map books I use most are from the DeLorme Company of Freeport, Maine. DeLorme makes map books of entire states, and the maps are good highway references. In most cases, DeLorme maps are done topographically, which can be of enormous help in understanding the terrain of a planned outing or studying the environment of a species or site.

I would also suggest making your own site-specific maps when possible. In almost all cases of orchid sites, I do a sketch map first. I note any nearby landmarks, road numbers or names, compass direction, and mileages. I then use a drawing program to draft and store an enhanced map on my computer. You do not have to be an expert cartographer to draw good maps. Try to draw the map to include all the features that you would want to know if someone else were drawing the map for you. I also write down a description of the directions to a site and enter them separately in the computer. That way, I have two different types of information that I can retrieve quickly. And having data handy in this way is particularly valuable if you want or need to send directions to someone else—perhaps in trade for information on a separate site. I simply print out the map and directions and then make copies as desired. Using the computer is also great for being able to update files easily in case a road is altered, for example, or you discover other new information about a site.

As I have gotten older, I no longer simply get out and walk the mountains for days or even hours at a time as I once did. Now I use a method of exploration that I call the park-and-walk. I park in what I perceive as a good orchid area and simply try to survey that specific area thoroughly. If I am with friends and we have two vehicles, we choose a stretch of woods or roadside and leave one car at the end of our planned walk. If, as is often the case, there is only one vehicle and only one person—me—I park and walk one side of the road for a period of time or a measure of distance and then cross over and walk the other side back to my car. This might be considered by some as the lazy way of doing things, and it probably is. But, primarily because of exposure to better light conditions, far more orchids are found along the edges of woods, roadsides, and trails than are found deep into any woodland.

Master the use of a simple instrument called a compass. When entering unfamiliar terrain, you should always have a proper plan of exploration, and a compass can help you with your plan. On a more technologically advanced level, one of my friends has started experimenting with the new global positioning system (GPS). This handheld, calculatorlike device can pinpoint your location using signals from satellites that indicate the latitude and longitude. This can be extremely helpful in finding your way and in relocating a plant site should returning to the same spot otherwise prove difficult. One drawback of the GPS is that it does not work well on cloudy days or underneath a heavy tree canopy. So approximations may be necessary.

With experience, it becomes easier to recognize certain areas that are more likely to have particular orchids. Learn to identify other wildflowers, trees, and shrubs that tend to share the same habitat as the sought-after orchid. An introduction to the basics of geology can also be of tremendous help in learning about orchids. Knowing which types of vegetation and which kinds of rock indicate acidic or more basic soils can make the orchid pursuit easier and much more fun.

One last bit of advice is to use binoculars. My wife and I are avid bird-watchers, so naturally we always try to have binoculars with us, whether we are driving or hiking. I have found that most people who enjoy birds also have a fondness for wildflowers. But many do not recognize the enormous help that binoculars can offer in spotting wildflowers. You would be surprised at how much tall phlox looks like purple fringeless orchid in a late-summer meadow. The two plants are the same color, the same height, and the same general shape. Binoculars can make the difference in determining whether or not you walk out into the meadow only to come upon a healthy stand of phlox. Time after time, when I have spotted a plant high on a bank, well past the "telling for sure" point, binoculars have saved me from having to scramble up the bank to try and determine its species. As the years go by, I appreciate more than ever the steps saved by my trusty binoculars.

Hunting wild orchids is not difficult at all, but finding them sometimes can be. Learn the tricks of the trade, and orchid hunting and finding can be fun as well as educational; and it can create treasured memories that are valued forever. Find, and consult, people who know about wild orchids. Learn where to look for wild orchids, when to look for the particular species you seek, and how to look effectively. And never give up on the resolve to find whatever might be around that next bend in the trail.

Special Orchid Places in the Southern Appalachians

In the southern Appalachian Mountains there are many special orchid places, far too many to list here. But some special places that consistently produce a variety of beautiful wild orchids cannot go without acknowledgment. Some of these places have well-known names, while others have names that are less well known. Some of the very best have no particular name at all.

The Cranberry Glades

The orchid place with the most recognized name in the southern mountains is the Cranberry Glades in Pocahontas County, West Virginia. Administered by the Monongahela National Forest, the glades are only one part of a vast botanical wonderland. A fine visitors' center for the area is located along Highway 39 just a few miles west of Mill Point at the intersection with U.S. Route 219. There are five glades altogether, and a boardwalk traverses two of them. The glades are northern-type bogs that began to form when water was trapped in this high mountain valley after the last southward advance of the glacial ice. As vegetation has slowly crept in replacing the water, deep accumulations of decayed vegetative material have been deposited in the form of peat (the beginning of the process that leads to the eventual formation of coal). In Canada, bogs are also sometimes called muskeg. True muskeg is a "trembling earth" environment: one can literally bounce up and down on the surface and the entire area will begin to undulate.

Orchids begin blooming at the glades in late May and early June with the appearance of heart-leaved twayblade and northern coralroot orchid. Although these species are common in the North, in the southern Appalachians they are known to grow in only a handful of sites. The early variety of the spotted coralroot orchid is next to be found in and about the glades. The orchids are at their peak in late June and early July with the coming of the true bog orchids, like grass-pink orchid and rose pogonia. Both of these species are readily seen from the boardwalk, and hundreds of plants of each species can be seen in a normal year. Mid-summer is the domain of the fringed orchids, and the Cranberry Glades have more than their share. The finest of this group is the hybrid known as Keenan's fringed orchid. The orchid season at the glades extends into early November, when both the nodding ladies' tresses and the more uncommon yellow ladies' tresses are easily located there. No less than 16 species of native orchids make their home in the Cranberry Glades.

Down the Blue Ridge

In North Carolina and Virginia, another special place is the Blue Ridge Parkway. Begun in 1935 but not totally completed until 1987, this ribbon of asphalt connects Shenandoah National Park in northern Virginia with the Great Smoky Mountains National Park some 464 miles to the south in North Carolina. Every mile is marked with a concrete milepost that wonderfully provides a place of reference for the wildflower lover who wants to relocate the area of a noteworthy find.

The wildflower season starts early in April at the lower elevations of the Blue Ridge Parkway. But much of the parkway and its botanical treasures are at high elevation, especially the areas around Asheville, North Carolina. The tremendous range of elevation gives the wildflower hunter a unique opportunity to prolong the blooming season. One can easily find places to enjoy spring flowers in the lowlands in early April and then go to the high country and find places in which to see the same species in bloom again several weeks later. Rhododendron puts on a spectacular show along much of the parkway in late May and early June, but it is late June before the performance is repeated at high elevations such as Craggy Gardens in Buncombe County, North Carolina.

Dozens of wild orchid species are found all along the Blue Ridge Parkway, much of which winds through wilderness and mature woodlands, making it a perfect route to prime territory that would otherwise be much less accessible. Being restricted by the narrow park boundaries is no problem for the orchid seeker. Roadside botany, in this case, is not just the best way but about the only way. In mid- to late May, both pink and yellow lady's slippers are found in abundant numbers and are sometimes visible from the roadside. With persistence, in early June one can find two tiny-flowered species of orchid, lily-leaved twayblade and Loesel's twayblade. Small purple fringed orchid and large purple fringed orchid can easily be seen in late June, sometimes side by side along the edge of the parkway near Mount Mitchell in North Carolina. Both of these orchid species are exceptionally beautiful and cannot help but catch the eye of the careful observer.

Dark hemlock or pine woods along the parkway often reveal the circumboreal lesser rattlesnake plantain, a rare orchid with an engaging rosette of checkered leaves and a short spike of tiny white flowers clinging to only one side of the stem. But the meadows as well as the woods are worth a look along this unrivaled scenic drive. They provide habitat for a number of species of the fringed orchids, such as yellow fringed orchid and ragged fringed orchid. The meadows also contain a number of species of ladies' tresses, some common and some, like spring ladies' tresses, quite uncommon. I know also of several meadows that have populations of the tiny and very difficult-to-see green adder's mouth orchid.

Canada in Virginia   The Mount Rogers National Recreation Area in southwest Virginia, near the North Carolina border, is a high-elevation area. Red spruce and Fraser fir forests on the ridge tops and beech-maple slopes beneath give the area the appearance of a chunk of Canada that has been dropped off in the southern Appalachians. On Mount Rogers itself, the elevation reaches 5,729 feet, the highest point in Virginia. But Mount Rogers also has broad shoulders reaching out to the east, west, and south. Wilburn Ridge, one of these shoulders to the south, is a wide-open expanse of land called a bald.

Balds are high-elevation, open grassy meadows exposed to extremes of harsh weather that cause a natural bonsai effect as cold temperatures and a constant wind do their own cropping of the vegetation. Wilburn Ridge is accessible by the Appalachian Trail and stretches south to

Massie Gap in the adjoining Grayson-Highlands State Park. Like so many of the southern mountain balds, the Mount Rogers bald area is wilderness and demands respect. The unpredictable sudden weather changes can be brutal at this high elevation. People have been known to lose their way here in the frequent, bone-chilling fogs or winter whiteouts and die as a result.

The huge open meadow at Massie Gap is easily accessible and is a veritable garden itself. The panoramic backdrop of gray outcropping rhyolite boulders that dominates the horizon is every bit as enchanting as any view in Colorado or Wyoming. Tens of thousands of high-bush huckleberry plants are scattered over the hundreds of acres of open bald. In this meadow is a special place where I have found eight species of wild orchids.

Pink lady's slipper begins the orchid parade at Massie Gap in June, soon followed by Small's twayblade and the small purple fringed orchid. I also once found here the very uncommon hybrid, Andrews's fringed orchid, which blooms in late June and early July. The club-spur orchid and the green adder's mouth orchid bloom in the meadow from late July into early August. In September, the orchid season bows out at Massie Gap with a brilliant display of nodding ladies' tresses. This meadow will delight the heart of anyone who appreciates majestic scenery and especially anyone who appreciates wild orchids. In the years before the state park was officially opened, I made excursions to this meadow many times, often alone. In those days, I would sometimes spend an entire weekend hiking and enjoying nature without encountering another human being. Now that Grayson-Highlands State Park has become a popular recreational area, with easy access to the Appalachian Trail and the wilderness area beyond, finding a parking place among the rows of cars at Massie Gap on a summer weekend can be a challenge. But the orchids persist, and this meadow at Massie Gap remains a genuinely special place.

The Great Smoky Mountains

Of course, no discussion of the southern Appalachians would be complete without mentioning the Great Smoky Mountains National Park. Located roughly east of Knoxville, Tennessee, and west of Asheville, North Carolina, along the border between the two states, this impressive range of mountains includes some of the more challenging wilderness in the East. Yet the area is readily accessible from Interstate 40. This rugged land offers so many types of environment that it is a naturalist's dream. Whether one is interested in wildflowers, mammals, insects, salamanders, geology or almost any natural phenomenon, these mountains have something to offer. The mountains are also a special orchid place, with more than 20 species, some rare and some common, within the boundaries of the national park.

Orchids span the flowering season in the Great Smokies. In mid-April, showy orchis is prolific in the lowlands on the Tennessee side of the mountains, providing a real treat for early spring visitors to such areas as Little River Road, Greenbriar Cove, and Noisy Creek Trail. In May, puttyroot orchid and yellow lady's slippers can be found at middle elevations just off the highway near Cade's Cove. On the North Carolina side, Heintooga Overlook Road, in a high-elevation area, often has hundreds of gorgeous small purple fringed orchid plants along its wet banks in late June, and the same area has yellow fringed orchid in August. I once found several plants of the secretive crane-fly orchid in the dark woods near Indian Creek Falls in the Deep Creek section of the park south of Cherokee.

The orchids are only part of the botanical treasures of the Great Smokies. Guided walks at the wildflower pilgrimage in late April each year can introduce newcomers to the fantastic flora and fauna of the area. From low-elevation coves hidden away along rushing mountain streams in hardwood valleys to high-elevation spruce forests, the Great Smokies have areas that can be enjoyed year round. A widely diverse topography located in a temperate climate zone has provided these mountains with one of the world's most richly diversified floras.

Unglorified Special Orchid Places

From their natural geographic placement, some special locales just seem to be havens for especially rare orchid species. Time and again, when particularly uncommon orchids appear in the southern Appalachians, they seem to turn up in certain areas with a high rate of frequency. In Virginia, this area is Augusta County, located in the west central part of the state. In southwestern North Carolina, the prize goes to Henderson County, located on the eastern edge of the Blue Ridge.

One of the extremely rare species of native orchids found in the southern Appalachians is the eastern prairie fringed orchid. As its name implies, this is a plant of the prairies of the upper Midwest. Far outside its normal range, this orchid is found in only one spot in the southern mountains, a wet prairie in Augusta County, Virginia. A prairie habitat in Virginia is extraordinary to begin with, but the presence of this exceptional orchid at this location makes it a doubly special place. In a place called Magnolia Swamp, Augusta County is also home to one of only two locations in Virginia where bog rose or arethusa has ever been found. In fact, of the 48 species of orchids recognized by the *Atlas of the Virginia Flora* as occurring in the state, 31 have at one time or another been present in Augusta County.

North Carolina's answer to Augusta County is Henderson County. It is one of only three counties in North Carolina where the exceptionally rare small whorled pogonia has been recorded and one of only three counties where collections of bog rose have been made. It is the site of North Carolina's only mountain location for the white fringed orchid and the crested fringed orchid.

The westernmost extension of the southern Appalachians is the Cumberland mountain range of southeastern Kentucky and eastern Tennessee. As the Cumberland Mountains extend to the southwest, the area becomes the Cumberland Plateau. The northern part of the plateau is defined as the area whose drainage goes into the Cumberland River. The winding course of the Tennessee River clearly defines the eastern edge of the plateau to the southwest of the city of Knoxville, and the southern boundary is defined by the high ridges to the west and north of the city of Chattanooga. The western outline of the Cumberland Plateau is a circuitous meandering of ridges that become rolling hills as the land tilts westward toward the city of Nashville.

The geologic significance of the Cumberland Plateau as it affects the native orchid population of the southern Appalachians was noted in the Introduction. The conducive conditions of the plateau ensure that several notable orchid species are found there. The shadow witch orchid, ordinarily a coastal plain species, makes its only appearance in the mountains on the western edge of the plateau in southern Tennessee. There are only a handful of locations in the world for monkey face or white fringeless orchid, and almost all of them are found on the Cumberland Plateau in Tennessee. Normally coastal plain species such as crested fringed orchid and yellow fringeless orchid appear on the Cumberland Plateau. Both Kentucky and Tennessee boast approximately 30 species of native orchids just on the Cumberland Plateau. Many of the area's deep ravines and hollows still have not been botanized, as evidenced by several relatively recent discoveries of the Kentucky lady's slipper there. I personally found a new location for this orchid along the Big South Fork of the Cumberland River in Tennessee in 1996. Who knows how many more populations of this plant and other rare orchid species are to be discovered in the rough terrain of southeastern Kentucky and northeastern Tennessee.

The northern Cumberlands are a rugged mountain range of narrow valleys and steep ridges where the sun rises late and sets early. This is a land of coal, dry shaly hillsides, copperheads, and hard times. It is also a land of autumn color, spring wildflowers, sparkling streams, patchwork quilts, and God-fearing neighbors. And there are also significant orchid populations in the northern Cumberlands. Pine Mountain, a formidable northeast-to-southwest-spreading ridge that divides parts of Kentucky and Virginia and extends well into Tennessee, harbors the gorgeous mountain species now called smaller rosebud orchid and the typically southern crested coralroot or cock's comb orchid.

Some of my personal favorite orchid places have no name at all. In Giles County, Virginia, which adjoins my home county of Pulaski, there are two particular roads that slice along two different creeks running through the Jefferson National Forest, one in the eastern part of the county along Big Stony Creek and the other along Dismal Creek in the western part. Each area is traversed by the Appalachian Trail, and each is home to black bear, white-tailed deer, and wild turkey. One is a road of dust or mud, as the weather dictates. One is a paved road. Both are crowded with trout fishermen in the spring wildflower season and both are abundantly supplied with orchids.

Collectively, on these two stretches of roadway I have seen 29 species of native orchids. Every single one of them has been near the roadside. Included in this list is my own discovery of a location for showy lady's slipper, an event I will recall fondly for the rest of my days. Near a northern bog habitat, the paved road has yielded a rare site for shining ladies' tresses, a typically northern species. Here too is the world's only location for Bentley's coralroot orchid. Orchids can be found along these two roads from spring into autumn. These are unquestionably special places for all seasons.

Sometimes, special places are simply very small plots with no particular distinction other than the fact that orchids grow there. I recall a hillside in my home county where I once found 117 pink lady's slipper plants blooming in a circle that I roughly measured to be 17 feet in diameter. On another bank along the Blue Ridge Parkway, I found more than a hundred plants of large whorled pogonia growing in a single tight clump, which, to my pleasant surprise, proved to be just beneath several dozen plants of pink lady's slipper. Once, while walking the Appalachian Trail, I came upon a cluster of 37 yellow lady's slipper plants in a grouping no larger than an automobile steering wheel—and, to top that, all but two of the plants were double-flowered.

Trying to list my special orchid places, even just a few of the best, would be a task I could never complete and would never want to complete. My personal special places are not in all instances filled with great numbers of orchids or necessarily blessed with exceptional beauty. The really special places are preserved like small parcels in my memory. Some habitats that I have known have now been lost. Some have changed, for better or worse. But my memory of these places and the joy they have brought persists—and, with luck, I will encounter many new places to call special.

Preserving Wild Orchids and Their Habitat

I have always closely guarded the location of orchid sites, and I was wary of providing such directions as I have in some instances in this book. But I have tried to be sure that any such information about the location of orchid places is about sites that are on either federal or state property and are protected by appropriate laws. It is no secret that people are prone to dig and try to transplant orchids, often with disastrous results. Both special knowledge and special habitat are required in order for transplanted orchids to prosper. The transplanting of native orchids from the wild to the home garden should never be attempted in today's world. Considering the constant assault on the habitat of our native plants in modern times, we cannot tolerate deliberate removal of our native orchids. Injury to our wild orchid population—or any other floral population, for that matter—is one from which we may never recover.

Even responsible collecting can alter orchid populations. Some orchid species are so common that one might think little of collecting a specimen; but, if we can compare plants and animals, so the passenger pigeon was once a very common bird species. In any event, it is seldom common species that are sought for collecting. It is rare species or species found well out of their normal range that bare the brunt of the pressure from collectors. Legitimate and selective collecting of plant materials has been absolutely necessary to the advancement of botanical knowledge. I emphasize *has been*. Today we have still cameras, video cameras, and digital cameras connected to computers that can supply almost any information needed for detailed and diagnostic taxonomic classification. With the technology that is already available, and advancements arriving almost daily, I hope that there will soon be a time when herbaria will at least be able to limit drastically their need for a "hard copy." The computer age is here; do we need to remove every plant specimen from the wild when we can gather, store, and exchange the information electronically?

The Blue Ridge Parkway under Siege

No native orchid should be considered so abundant that it does not need protection. Our history of decimating other species (both animal and plant) has made that lesson very clear. When habitat is destroyed and plants are lost by the hundreds, a parcel here and a parcel there, eventually even the most common of plants will feel the impact of lost habitat. And those of us who care about these plants and these habitats will suffer a loss that could never be regained. This imminently irretrievable loss of habitat makes the existence of organizations and preserves directly involved in keeping the land natural extremely important. We can promote orchid preservation by supporting these efforts. There are many such groups and preserves that do a wonderful job as stewards of the land. A group such as the Nature Conservancy uses its resources directly to buy land and ensure its preservation. But, such private efforts notwithstanding, we rely on federal and state governments to protect the majority of our natural history. This government protection takes many forms: wildlife refuges, parks, forests, recreation areas, and others. In the eastern United States, the Blue Ridge Parkway is a primary example of our government's efforts. It has served not only to protect habitats and species but also to preserve an insight into an isolated mountain culture that, in many respects, never kept full pace with the rapidly changing world around it.

The Blue Ridge Parkway extends from Rockfish Gap on the border of Augusta and Nelson Counties, Virginia, southward to just north of Cherokee, North Carolina. Building this linear park was a massive undertaking both physically and environmentally. As it passes by the highest elevation in eastern North America at Mount Mitchell and breaches the high country south of Asheville, or as it crosses the low farmland along the Roanoke River, the Blue Ridge Parkway continues to be a wonderland of opportunity for anyone who enjoys any part of being outdoors. It provides an opportunity matched by few places anywhere in the world.

Yet, today, the parkway is under attack from many sides. Private landholders and real estate entrepreneurs have begun to recognize the prestige and financial benefit that can be derived from development along the parkway. Housing developments have begun to spring up on the ridge sides, even in places where one might think it impractical if not impossible to build a house. The once breathtaking views are spoiled beyond repair. Almost every time I visit the parkway these days, I see a new building. As I point my camera toward the horizon, I find it increasingly difficult to take a picture that doesn't include a microwave relay antenna or a new log home. Just recently, it was announced that a large development that would be in full site of parkway visitors would be built near the parkway east of Roanoke. There was a public outcry, but, after all, it was private land. The developers did finally hear the protests and agreed to some modifications, planting trees to block the view from the parkway. But the houses are still being built and winter still comes and leaves still fall, and this once wild part of the Blue Ridge Parkway will still be lost forever. Where next?

As one heads north on the Blue Ridge Parkway from the Great Smoky Mountains, the highest elevations of the entire parkway are soon encountered. This area is a prime example of a Canadian-zone wilderness in the eastern United States. Until relatively recently it was a pristine place where (even into the 1970s) one could look back to the southwest and enjoy, unimpeded, the wonderfully brooding silhouettes of the Great Smokies. But now this landscape is dotted with a miserable patchwork of clear-cut logging scars. It is virtually impossible to use a wide-angle lens here without including these unsightly gaps in the once singular mass of forested mountains. More and more each year, the Blue Ridge Parkway loses a part of its once untamed enchantment. What will be left for our grandchildren?

The Preservation Dilemma

Since the federal and state governments are the largest landholders in the southern Appalachians, a large percentage of orchid sites are on "public" property. When it comes to dealing with exceptionally rare species on public land, there are a number of concerns that the administrative agencies of the government have to take into account. Decisions made will affect all of us and the future of plant conservation.

The example of one particular site for the exceptionally rare and endangered small whorled pogonia in southwest Virginia has given me a renewed appreciation of the complexity of conservation problems and respect for officials who must deal with the public while, at the same time, making the best possible decisions for the plants. The site I refer to is on national forest land and, due to the restrictive policy of the forest service, the discovery of the site was not publicized. Only a very privileged few even knew about it.

When I finally heard about the site and tried to see it, I had considerable difficulty arranging a visit. Through the cooperation of Fred Huber, forest ecologist for the George Washington and Jefferson National Forests, I was finally allowed to visit the site in 1996. My primary interest was to compare the habitat of this site with that of others of which I had knowledge. To me the knowledge gained from experiencing this site had immense implications.

The habitat was unlike any of the sites I had seen in New England. This site and one other that I later visited in Georgia completely reshaped my thinking about the habitat that is required for the small whorled pogonia. I know now that this special orchid is not restricted to a single type of environment. This information has been critical in the formulation of my thoughts for this book. I consider this a most legitimate use of knowledge achieved through cooperation with the United States Forest Service. But what about other people with different agendas who consider their agendas just as important as mine? Will they have the same opportunity I had? Is fear of "diggers" or vandals legitimate enough to keep all people away? The forest does belong to the public.

The forest service is now faced with a perplexing dilemma. With the growing number of known populations of small whorled pogonia and other endangered plants comes a widening awareness among botanists and a growing demand for information. And, of course, an increasing number of folks want to see the plants. Deciding who and how many are allowed access is an unsettling problem. It is a tough decision. How is the decision made? How can the sincerity and truthfulness of interested people be measured? As of this writing, it seems that forestry officials are taking a very conservative approach. In the spring of 1996, a group from the Georgia Native Plant Society was scheduled to make a trip to see a site for small whorled pogonia in the Chattahoochee National Forest of north Georgia. But just a few days prior to the scheduled date, the forest service denied them access, saying they were unsure about the impact so many visitors would have on the plants. Information from many areas and many botanical sources has made it clear that the small whorled pogonia is a plant that does need protection, and a great deal of it. But ultimately one must ask the question, for whom are we saving the plants? If not for the people whose tax dollars support our public lands, then whom? Yet, we must also ask whether we are saving the plant for the sake of people or for the sake of the plant itself. These are complex questions that merit careful consideration.

Nature's Balance

In August of 1976, I found a particularly nice population of yellow fringed orchids in my home county of Pulaski in

Virginia. There were more than 200 flowering plants in an area that had been clear-cut many years previous. The canopy of young trees was almost complete, and the flowering racemes on the orchid plants were all relatively small, both indications that this population was probably in decline. Sure enough, within seven years, the population of orchids at this site was zero. Before the cutting, the shade from the mature trees probably did not allow the growth of any yellow fringed orchid plants at this site since these orchids require a great deal of sunlight. After the harvesting of the trees, the orchids were able to "move" in and flourish, taking advantage of the open sunshine. But after the habitat again changed, the plants could no longer thrive.

Another yellow fringed orchid site, in North Carolina, had more than 700 plants in bloom when I found it in 1992. When I revisited it a mere two years later, there was only a handful of plants scattered about the edges of the former colony. A highway department ditching project had not only destroyed most of the plants but also greatly altered the drainage of the area so that no water—water needed by the yellow fringed orchid plants—settled in the habitat. Habitat had been destroyed and plants lost.

Many plants and animals tend to be very sensitive to their immediate environment, their habitat. Sometimes even a minuscule change can affect the stability of the population within the area. So it is important that we look at plant preservation as habitat preservation. Surely, if we continue the course of habitat destruction we have undertaken, the destruction of the plants cannot be far behind.

There must always be a place for each species, a place where that plant or animal can survive in its optimum numbers for the optimum length of time. This is what is known as the balance of nature. With any interruption in this balance, this cycle of life, some species will suffer or disappear altogether somewhere along the way. Nature has worked to maintain the equilibrium of this balance over the millennia. Humans have been the primary interlopers who have thrown many life cycles into utter chaos and totally destroyed others. Yet, as the dominant species, mankind remains the greatest hope against further spoilage of our natural heritage and for the restoration of habitats that are still capable of being saved. Only humans can legislate for and bring about real conservation and preservation of our natural environment, and only they can see this world as one global community, interdependent with all of nature.

The native orchids are only a part of the whole, just as all things—ourselves included—are a part of the whole. And, some would argue, our future can be measured by gauging the way we treat the smallest of things. Only by educating ourselves on the wisest uses of our resources can we hope to avoid the mistakes of those who have preceded us. Only by recognizing what must be done and by being willing to make the sacrifices necessary can we save our children from having no forests in which to walk, no songbirds to greet the dawn, no flowers to smile up at the sky. Native Americans view life as a circle, every part connected. If we lose a part of our world around us, we lose a part of ourselves. If we lose respect and simply cast away any section of the circle no matter how tiny, there is no more circle and we begin to spiral out of control toward our own extinction. Like all species, humans have evolved as they have only because they found a suitable niche in which to exist and a renewable food supply to sustain them. We are not the circle, only a part of the circle.

Using This Book

This book is a guide to the orchids that can be found growing wild in the southern Appalachian Mountains. It is not intended to be a technical work used only for reference by botanists or rabid orchid lovers. Rather, it is intended to be used and enjoyed by everyone, from those who simply enjoy photographs of beautiful wildflowers to those who are professionally involved with botany. The primary purpose is to illustrate the unmatched beauty and diversity of the native orchids found in our southern mountains. This book is an attempt to share the boundless joy and delight that have been mine over nearly three decades in the pursuit of orchids. I hope it will contribute to a better understanding of our natural heritage and expand our appreciation of the wonder of nature in general and the treasures with which we are blessed in the southern mountains in particular.

The native orchids of the southern Appalachian Mountains have not heretofore been treated comprehensively in book form as a regional entity. Oscar W. Gupton and Fred C. Swope's *Wild Orchids of the Middle Atlantic States* (1986), covered the entirety of eight states and included the area of the southern Appalachians but on a much more general level. Since that time, there have been more than a half dozen species added to the orchid flora of the southern Appalachian Mountains alone. There also have been numerous nomenclatural changes and reclassifications of the species, varieties, and forms of the orchids of the southern mountains. This book incorporates such new information and provides a comprehensive and up-to-date survey of the native orchids of the southern Appalachian region.

It seems wise to offer here a few words of explanation about the names of orchids since those names will be important for the discussion of various species in the remainder of the book. Like all other plants, orchids have a common name and a scientific name. Common names for orchids can vary widely from one region to another and from one generation to another. There are no standards for the assignment of common names and thus such names may often be confusing.

Scientific names, though often difficult to pronounce or remember, are more precise. Carolus Linnaeus, an eighteenth-century Swedish botanist, refined an existing but little-used system in which all plants have a two-word scientific name. This system is known today as the Binomial System of Nomenclature and is regulated by the International Code of Nomenclature. As this two-word naming system has evolved since Linnaeus's day, certain universal rules of nomenclature have been recognized. Among these rules are the standard that the name is intended to be the same worldwide and that the first name legitimately applied to a plant (i.e., published as a name for it) should take precedence over all others.

Within a plant family, plants that are very similar and can fit into the same group are first given a name according to genus (pl. genera). Then a specific name (species) indicating the plant's individual characteristics within the genus is added. The specific name can be descriptive, the name of a person, the name of an area, or some other name that would set that species apart from others. The specific name is "Latinized" to conform to scientific rules. The genus name is always capitalized and the species name lowercased, and both names are italicized in print. Although the genus and species names are often used alone for brevity's sake, the scientific name is not really complete unless the genus-species combination is followed by the name or initial of the authority who first gave that name to the species in question. Because of his well-known work in developing thousands of plant names, most of which remain in use, a simple "L." is understood worldwide to designate Linnaeus as the cited authority.

After having once been given in full within a single text, the scientific name can be slightly altered: when repeated, the genus may be abbreviated with a single capital letter and a period. For instance, the scientific name for pink lady's slipper is written as *Cypripedium acaule* Aiton. When repeated within the same text, the name can be written *C. acaule* Aiton (or, in shortest form, *C. acaule*).

In this book, 52 species of native orchids are described within 21 genera. (See the beginning of Part 2 for an overview list of the genera covered.) Of this number, one has been given only cursory consideration. Helleborine orchid is an alien species imported from Europe that has spread across much of the eastern United States. Three native hybrid orchids are given full attention: Andrews's fringed orchid, Keenan's fringed orchid, and Jones's twayblade. To designate a hybrid in scientific terms, a hybrid sign (×) is placed between the genus and the hybrid name with no space between the × and the hybrid name. For instance, Keenan's fringed orchid is known scientifically as *Platanthera ×keenanii* P. M. Brown (P. M. Brown being the botanist or authority who assigned the name).

"Variety" is a formal botanical term used to describe definite differences within a species that are present but are not pronounced enough to differentiate the plants exhibiting the differences as a separate species. Some of the orchid species in the southern Appalachians exhibiting varietal differences are the yellow lady's slipper and the pad-leaf orchid, both of which have a large and a small variety; the tubercled orchid and the slender ladies' tresses, each of which has a northern and a southern variety; and the spotted coralroot orchid, which has a distinct western, or early, variety. To designate variety, the abbreviation "var." or simply "v." is used, as in *Cypripedium parviflorum* var. *parviflorum* to designate the small variety of the yellow lady's slipper.

A number of species have recognized forms that generally imply color diversity, such as the so-called albinos. Forms are usually written as "forma," for example, *Corallorhiza maculata* forma *flavida* to designate the yellow form of spotted coralroot orchid.

One species, green adder's mouth orchid, is treated in this book as a single species although orchids bearing that name are recognized by several authorities as including two separate species. All of these hybrids, varieties, and forms are given full explanations in conjunction with the species descriptions in this book.

The individual species of orchids described in the book are grouped by genus, with the genera arranged alphabetically according to scientific name. When a genus contains only one species, the species description also describes the genus, of course. In cases where several species are present in a particular genus in North America, in or out of the southern mountains, a separate section describing the genus precedes the species descriptions. The individual species are then listed alphabetically within the genus according to their scientific names. The heading for each species also includes its most often used common names. In some cases, other commonly used names are described in the text of the species descriptions. I believe that listing the species within their own genera rather than by color or some other characteristic, is the best arrangement for comparing similar species.

Descriptions for each species include (in roughly this order) information on the following: the plant itself; the flowers; the time of flowering in the southern mountains; the range, both in the southern Appalachians and across North America; its relative abundance; and its habitat within the southern mountains.

I have purposely avoided the use of scientific names for species, orchid or otherwise, mentioned in Part 1 of the book. Scientific names are more appropriately applied, however, in the species profiles. But even there, whenever everyday terminology is practical, I have simplified the text by avoiding scientific names or descriptions of the orchids. Generally, proper scientific names are given only once within each separate species description. Still, the scientific names are important and are provided in this book not only for botanical correctness but also for the reference of readers who wish to seek further information about any of the orchids in other works that may include more scientific data than is presented here. The large majority of the scientific names I have used are taken from Carlyle Luer's *Native Orchids of the United States and Canada* (1975). Scientific names taken from other sources have primarily been taken from authorities published after 1975. In particular, the Kentucky lady's slipper's name, *Cypripedium kentuckiense*, although proposed earlier, was established by C. F. Reed in articles published in 1981 and 1982. This scientific name has become the one most used in today's orchid literature. Also the splitting of rosebud orchid, *Cleistes divaricata*, into two separate species by Paul M. Catling and Katharine B. Gregg in 1992 warranted a name for the smaller mountain species. The smaller rosebud orchid's scientific name became *Cleistes bifaria* and is used in this book. The North American yellow lady's slipper has been formally named *Cypripedium parviflorum* to help separate the plants occurring here from the European yellow lady's slipper traditionally named *C. calceolus*. From the southern Appalachians, the addition of Bentley's coralroot orchid, *Corallorhiza bentleyi*, places an entirely new species in the orchid family.

The species descriptions are designed to provide clear information for readers using them in order to identify plants in the field. In certain cases—particularly the ladies' tresses, for example—descriptions are difficult and even trained botanists often have trouble distinguishing some of the species in the field. But I have always tried to give the most obvious field characters that would help in identification. Color, shape, and size are generally the most important of these. Native orchid flowers vary in color from the most drab greens to the most spectacular magentas and yellows. The shape of the flowers varies from the obviously inflated pouch of the lady's slippers to the insectlike blooms of the tiny twayblades. The size of orchid flowers ranges from less than a quarter inch across in some instances, such as the green adder's mouth orchid, to 3 inches or more in width in some of the lady's slippers.

The species discussed are generally shown in at least two photographs. One shows the flower of that particular species so it can be distinguished from others. Another photograph displays the entire plant, as often as possible in the wild amid its immediate habitat. All photographs are by the author.

For each of the 52 species, I have provided a map that illustrates the known range, by county, across the southern Appalachian Mountains. Generally, there are no maps included for hybrids or forms. However, because their ranges are so widespread and more clear-cut than in other species, I do include maps showing the range of the northern and southern varieties of the tubercled orchid, *Platanthera flava*, and the slender ladies' tresses, *Spiranthes lacera*. A map is also provided for the western or early variety of the spotted coralroot orchid, *Corallorhiza maculata* var. *occidentalis*. Regarding the special cases of the yellow lady's slipper, *Cypripedium parviflorum*, and the pad-leaf orchid, *Platanthera orbiculata*, the range information for the southern Appalachians is incomplete, so a single map to indicate the range for the species as a whole is given, and no map has been provided to separate the known locations for the large and small varieties of these two species.

A large number of the county records represented on the maps are my own and have not been previously published, many simply because I chose not to disturb populations by taking specimens. In some cases, I thought the species to be too rare to disturb; in others, I thought it too commonly known to make much of an impact on collection data one way or the other. More simply stated, I just do not feel good about taking orchid specimens and do so only under rather strict guidelines. Bear in mind that the maps indicate only known locations for the plants. The simple fact that a certain area on a map is not indicated as a site for a particular species does not necessarily mean that the plant does not exist there, merely that no record has been formally reported for that area. The primary sources for these orchid records lie within the already published floras of each state involved. These records are supported by voucher specimens at several herbaria located across the Southeast and elsewhere. In particular, I have relied heavily on the following: for West Virginia, the records of Scott Shriver, Al Shriver, and Clete Smith of Pittsburgh, Pennsylvania, and Strasbaugh and Core, *Flora of West Virginia* (1970); for Kentucky, Ettman and McAdoo, *Kentucky Orchidaceae* (1978); for Tennessee, the records of Dennis Horn, Charles Wilson, and the late J. I. (Bus) Jones and Chester et al., *Atlas of Tennessee Vascular Plants* (1993); for North Carolina, Radford, Ahles, and Bell, *Manual of the Vascular Flora of the Carolinas*

(1968); and for Virginia, Harvill et al., *Atlas of the Virginia Flora* (1992).

For those interested in a more scientific description of native orchids, there are many good reference books in print. The most complete work on native orchids for all of North America is Carlyle Luer's *Native Orchids of the United States and Canada* (1975). The reader will find many references to Dr. Luer's work throughout this book. *Native Orchids of North America* by Donovan Correll (1950) is very informative and wonderfully illustrated with black and white drawings but is somewhat outdated now as to terminology. Correll's book provides range descriptions rather than maps. William Petrie's *Guide to Orchids of North America* (1981) is one of the most easily understood orchid books. *Orchids of the Western Great Lakes Region* by Frederick W. Case Jr. (1987) and *Orchids of Indiana* by Michael A. Homoya (1993) are both wonderful orchid books, and many of the species of the southern Appalachians are referenced in these guides. Many wildflower guides contain useful sections on native orchids. These works vary from the simplified Peterson Field Guide Series to the elaborate, multivolume *Wildflowers of the United States* by Harold William Rickett.

Throughout this book, references to North America are not meant to include Mexico or most of Florida, where expressly subtropical orchid species are encountered. References to the ranges of species, unless otherwise stated, should be presumed to refer to ranges within the southern Appalachian Mountains. Measurements given should be understood to indicate the size of typical plants within the southern mountains. Extremes in size, both large and small, have generally not been regarded.

Part Two Plant List

Native Orchid Genera of the Southern Appalachians

Scientific Name	Common Name
Aplectrum	Puttyroot orchid
Arethusa	Arethusa
Calopogon	Grass-pink orchids
Cleistes	Rosebud orchids
Coeloglossum	Green frog orchid
Corallorhiza	Coralroot orchids
Cypripedium	Lady's slippers
Epipactis	Helleborine orchids
Galearis	Showy orchis
Goodyera	Rattlesnake plantains
Hexalectris	Crested coralroots
Isotria	Whorled pogonias
Liparis	Twayblade orchids
Listera	True twayblade orchids
Malaxis	Adder's mouth orchids
Platanthera	Fringed orchids
Pogonia	Rose pogonia
Ponthieva	Shadow witch orchid
Spiranthes	Ladies' tresses
Tipularia	Crane-fly orchid
Triphora	Three-birds orchid

Opposite:
A typical plant of puttyroot orchid has a raceme of flowers superficially resembling that of the fringed orchids.

Left:
The puttyroot orchid flower does not have a spur.

The name *Aplectrum* is derived from words meaning "spurless." Although the flowers of puttyroot orchid plants are arranged in a raceme similar to those of members of the *Platanthera* genus (the fringed orchids), they lack the spur of the fringed orchids and also their bright coloring. The specific name, *hyemale*, is derived from a Latin word alluding to winter, which in this case refers to the winter leaf of the plant. This species is the lone representative of its genus in North America.

Puttyroot orchid flowers are dull yellowish or green in color and are very well camouflaged. The petals and sepals may be tipped with a dark maroon or purple color. The three-lobed lip is white with a few rather pale reddish-purple spots. The leafless stems are about 15–20 inches (approximately 38–50 centimeters) high. The stem and dried seed capsules are rather stout and often remain past the flowering time of the following season.

Aplectrum Nuttall

Aplectrum hyemale (Muhlenberg ex Willdenow) Nuttall

Puttyroot orchid or Adam and Eve

The hibernal or winter leaf of the puttyroot orchid is shown along with the new corm growing directly as an extension of the old one.

These puttyroot orchid "skeletons" standing in the woods sometimes offer an easy way to locate a new population from a distance. A variety of this species with pale apple green coloring is known from sites in Michigan, Ohio, and a few other areas.

Puttyroot orchid begins its blooming season in mid-April and, at that time, can be found along the one-way drive called the Roaring Fork Motor Nature Trail in the Great Smoky Mountains National Park in Tennessee. In southwest Virginia, these orchids wait until mid-May to flower.

The single winter leaf of the puttyroot orchid plant appears in early fall and remains green through the winter. The leaf usually, but not always, withers before bloom time. The leaf is about 2–3 inches (approximately 5–7 centimeters) wide and 8–10 inches (approximately 20–25 centimeters) long, with heavy parallel veining. This veining is often purplish or reddish in the young leaves of autumn but turns whitish in late winter. Looking for the winter leaves provides a much easier way to find puttyroot orchid plants than trying to pick them out of the shadows of the spring forest. The root, actually a corm, when cut open, has a puttylike consistency, thus the name. The other common name, Adam and Eve, comes from the fact that the previous year's corm does not completely go away before a new one is formed. The old, withering corm remains attached to the new one by a short rhizome in Adam and Eve fashion, the younger one springing from the older one, like Eve from Adam's rib.

Puttyroot orchid plants prefer rich, moist woods but are sometimes found along open roadsides or in sparsely wooded waste areas. This orchid species is found across the Midwest and over to the East Coast and from extreme southern Canada and mid–New England southward to the upper South. Puttyroot orchid is found fairly commonly over the entire area of the southern Appalachians.

Arethusa Linnaeus

Arethusa bulbosa Linnaeus

Arethusa or bog rose

In Greek mythology, Arethusa was a beautiful young woman pursued by the river god, Alpheus. Artemis, the Greek goddess who protected young maidens, tried to hide Arethusa by first turning her into a fountain and then into a hidden underground river. But Alpheus became a river himself and finally won his prize. Supposedly the two rivers now flow together near the coast of Sicily. No doubt the name of the genus of this orchid refers to the fact that the plant almost always finds its home in very wet surroundings. The specific name, *bulbosa*, refers to the corm or root, which is basically shaped like a bulb. Although over the years several other species have been listed under this genus, all have since been removed. In some areas, arethusa is commonly called bog rose, the name I prefer. And, because of its protrusive lip, this species is also sometimes referred to as dragon's mouth.

The typically singular flower of arethusa is a bright magenta color and sits atop a 4–6 inch (approximately 10–15 centimeter) stem. A single lanceolate leaf appears after flowering. The flower is small, about the size of the last joint on a human's thumb. The lip is curved downward like a gaping tongue with undulating edges and is white

Opposite:
Although small, arethusa or bog rose is among the most attractive native orchid flowers.

Left:
Arethusa plants are leafless at the time of flowering.

or pinkish with blotches of deep purple or maroon. There are fleshy, yellowish-orange bristles in the center of the lip that, in color and consistency, remind me of the pulp inside an orange. Arethusa in the southern mountains blooms in the first ten days of June and is restricted in habitat to wet, sphagnous bogs.

Arethusa becomes more plentiful as one travels northward. In areas of Quebec and Newfoundland, bog rose exists in large populations. Also, as one goes north, arethusa grows ever more variable in color. The flower color ranges, in some instances, from varying shades of the typical magenta to a form that is purple or bluish, and even an enchanting pure white form is not too difficult to find in the vast bogs of the far Northeast country. Based on color alone, some authorities have given form names to some of these variously colored plants.

Arethusa is the only representative of its genus in North

America and is particularly rare in the southern Appalachians. Arethusa has never been recorded in West Virginia, Kentucky, or Tennessee. Formerly there were two locations in Virginia, one each in Patrick and Augusta Counties. But no sightings have been made at either of these sites for many years, and the plant has to be considered extirpated in Virginia. One early collection was made in Henderson County, North Carolina, but this site is no longer considered viable.

Presently, only one small, fragile population of arethusa is known in the southern Appalachian Mountains. This site is a singular, tiny, very wet sphagnum bog in the Pink Beds of Transylvania County, North Carolina. The area is so named for the vast surrounding thickets of rhododendron and mountain laurel that bloom pink in late spring. The Pink Beds are part of the Pisgah National Forest, located near the Cradle of Forestry, the birthplace of America's first forestry school. Through the help of Rob Sutter, who was in the Pink Beds doing research on a rare member of the lily family named swamp pink (*Helonias bullata* L.), I first encountered bog rose in 1979. There were 17 plants blooming that year but only one flower the following year. My next visit was 17 years later in 1997. Unfortunately, on that visit, I could not find my way through the tangle of vegetation and was unable to locate the bog.

Considering the fact of a once much cooler, more northern climate and old collection records for former locations, arethusa likely once existed as a more prolific species in the southern Appalachians. Today, this is one of the handful of exceptionally rare orchids to be found in the area. Interestingly, except for the Augusta County, Virginia, location, all of the arethusa locations in the southern Appalachians have been found on the eastern slope of the Blue Ridge mountain range.

Calopogon R. Brown in Aiton

The grass-pink orchids

The generic name, *Calopogon*, is a combination of two Greek words that mean "beautiful beard" and refers to the yellow-tipped bristles or beard at the end of the lip. These bristles look as if they are laden with pollen, but the pollen is actually in pollinia, or pollen masses, on the end of the column, all the way on the other side of the flower.

Five representative species in this genus are known in

North America north of the subtropical zones. Three of these are confined to the coastal regions of the southeastern United States. A fourth species exists in the southern Midwest and was determined to be a separate species only in 1994. The species of grass-pink orchid found in the southern Appalachians has a much wider range than the other four. All five members of the genus have flowers that are similar in structure and in color, a rich, beautiful magenta. Sometimes white forms are also found in each species.

Calopogon tuberosus (L.) Britton, Sterns, and Poggenberg

Grass-pink orchid

The specific name for this orchid, *tuberosus*, is Latin and refers to the tuberous corms. Grass-pink orchid flowers are among our prettiest orchid species. The bright magenta-colored sepals and petals are large and very showy and stand in contrast to the bright yellow coloring of the beard that attracts pollinators to the lip.

Like the flowers of all species in this genus, grass-pink orchid flowers are nonresupinate (i.e., their buds do not make a 180° twist when opening, as do the buds of most other orchids). Thus, the lip remains uppermost. This unusual arrangement is an ingenious method by which the plant tricks would-be pollinators. The lip is connected to the flower on a hingelike structure. When the pollinator is attracted to the bright yellow bristles at the end of the lip and proceeds to land there, the weight causes this hinge to drop the lip. This brings the pollinator down so that it makes contact with the pollen grains on either side of the column. If there is already pollen stuck to the pollinator, that pollen is passed across the stigma to complete the pollination process. Pollinators must be of adequate weight, heavy enough to cause the lip to fall. Lightweights need not apply. When pollination is complete in a grass-pink orchid flower, the lip usually stays in a drooping position as if protecting the column, essentially closing the flower to further business.

Plants of the grass-pink orchid in the southern mountains have a thin stem that is typically 10–12 inches (approximately 25–30 centimeters) tall. Flower buds are large and round, with a pointed tip, and take on the magenta coloring well before opening. Luer stated that grass-pink orchid plants can have as many as 25 flowers. But in the southern mountains, it is unusual to see more

Opposite:
Its striking magenta color makes grass-pink orchid a favorite eastern wildflower. Note the uppermost, or non-resupinate, lip on the flowers.

Left:
Despite the appearance of the yellow-tipped bristles on the lip of grass-pink orchid, the pollen is on either side of the tip of the column opposite the lip.

than six flowers to a stem. I have seen as many as nine. Several narrow leaves are present, reaching about halfway up the stem. Plants begin to flower in mid-June in the lower Appalachians, but the prime time in prolific areas such as the Cranberry Glades is about the first of July.

Grass-pink orchid ranges across the eastern half of the United States to southern Canada and then northeastward through the Maritime Provinces to Newfoundland. This species, according to Luer, is also found in Cuba and the Bahama Islands. It grows in scattered locations throughout the southern Appalachians. It is very sparse in Kentucky, in areas along the Virginia border, but occurs at a number of sites in the mountains of southwest Virginia and southwestern North Carolina and in the Alleghenies of eastern West Virginia. The mountains of eastern Tennessee, which have many sites, are the prime area for this orchid in the southern mountains.

Grass-pink orchid is normally found in strongly acid situations in the southern mountains such as sphagnum bogs and grassy, wet meadows. A true bog plant, grass-pink orchid often grows alongside the large cranberry, *Vaccinium macrocarpon* Aiton. In most sites in the southern Appalachians, population numbers are generally low, so the plant is listed as uncommon. Locally, however, this orchid can be quite prolific, as it is, for example, at the Cranberry Glades and at Droop Mountain Bog, both in Pocahontas County, West Virginia.

Cleistes
L. C. Richard
The rosebud orchids

There are more than two dozen species in this genus, of which all but one, according to Luer, are restricted to South America. But the one North American species described by Luer in 1975 was divided formally into two separate species in 1992. Flowers in this genus are especially beautiful.

The word *Cleistes* is from the Greek "kleistos," which means "closed." This refers to the fact that the flowers are in a tubular shape, open only at the very end. In times gone by, the plants in this genus were included under the genus *Arethusa*.

Cleistes bifaria (Fernald) P. M. Catling & K. Gregg
Smaller rosebud orchid or smaller spreading pogonia

This species has been separated from the traditional classification, *C. divaricata*, only in the last few years. Previously, some authorities did list what they referred to as the mountain variety of rosebud orchid, which was called *C. divaricata* var. *bifaria*. This is the terminology that botanist M. L. Fernald chose to use when he revised *Gray's Manual of Botany* in the 1950s. By the 1990s, most botanists had formally recognized that there were two distinct varieties of the plant, one in the mountains and one in the coastal plain. Coastal plain plants were said generally to be larger, more deeply colored, and living in wetter habitat. Thus a natural distinction was often drawn between the coastal variety and the mountain variety. Later work in this genus showed, however, that the mountain variety was also present in the coastal plain.

Katharine Gregg, basing much of her work on a site in Barbour County, West Virginia, began to investigate the two varieties and, along with well-known Canadian or-

The delicate beauty of the smaller rosebud orchid is often hidden among the weeds and grasses of its habitat.

chid expert Paul M. Catling, studiously pored over collection records from all over the territories where both varieties had been found. In 1992 they published their findings, in which they named the former mountain variety *Cleistes bifaria* (Fernald) P. M. Catling & K. Gregg, the smaller rosebud orchid.

According to the recent work, the larger rosebud orchid is restricted to the coastal plain in areas from northern Florida to New Jersey. The smaller rosebud orchid is found in the coastal plain from North Carolina to Louisiana but is also the species of the mountains, the southern Appalachians. The word *bifaria* means "two ways" and is used here in reference to the fact that this species has two distinct habitats, the mountains and the coast.

The smaller rosebud orchid is one of the most impressive appearing orchids of the southern mountains. Plants are about 12–15 inches (approximately 30–38 centimeters)

tall, with a single leaf about halfway up the stem. The plant typically has a single tube-shaped flower about 2 inches (approximately 5 centimeters) in length. Occasionally, two-flowered stems have been seen.

The greenish-maroon brown- or brassy-colored sepals spread widely upward and serve to give this orchid an unmatched distinction. This spread appearance is the source of the other common name, spreading pogonia. The lion's share of this orchid's beauty lies within the flower tube, which can be either pink or white, formed by the forward-facing lateral petals and the lip. The lip protrudes beyond the other two petals and has a remarkably dazzling design of purple etching. The lip also has a few yellowish bristles that form a crest lengthwise along the center. A blotch of deep rose usually decorates the very tip of the lip. From this splash of rose, delicate purple veining radiates about the edges of the lengthwise crest in the fashion of narrow, jagged bolts of purple-rose lightning.

The smaller rosebud orchid blooms in early June in the southern part of the mountains but can be fresh in northern West Virginia as late as the first part of July.

Restricted to the southeastern United States, smaller rosebud orchid is very sparse throughout the southern Appalachians. It occurs at a few sites in eastern Kentucky as well as some scattered locations on the Cumberland Plateau and in the eastern mountains of Tennessee. It is infrequent in the North Carolina and Virginia mountains. And there are only two recorded sites for smaller rosebud orchid in the area of West Virginia covered by this book, one in Barbour County and one dating from 1968 in McDowell County.

All but one of the locations I have seen in the southern Appalachians for the smaller rosebud orchid are on well-drained, rather open, scrubby hillsides, often in open cuts of power lines. These areas have the typical ericaceous or heathlike vegetation of the majority of sandstone and shale regions of the Allegheny Mountains.

The Barbour County, West Virginia, site is a habitat of most notable exception. If anything, this site more resembles the coastal habitat than the typical mountain habitat for smaller rosebud orchid. The site consists of a wet, treeless meadow, and the plants are large and colorful. Bracken fern, *Pteridium aquilinum* (L.) Kuhn, regularly associated with dry areas, is prolific in the meadow, yet the soil is wet enough for crawfish holes and stands of

Opposite:
The loveliness of the flower of the smaller rosebud orchid rivals that of any orchid in North America.

cinnamon fern, *Osmunda cinnamomea* Linnaeus. The grasses are sparse and rather low in height. When viewing the site, I thought that if a few longleaf pine trees were added, the field would be almost identical to the savannahs of coastal North Carolina, where I have visited many sites for the rosebud orchid, both large and small.

Coeloglossum Hartman

Coeloglossum viride (L.) Hartman var. *virescens* (Muhlenberg) Luer

Green frog orchid

This orchid species, up until Luer's work in 1975, was usually included with the fringed orchids, which, at that time, were listed under the genus *Habenaria*, now *Platanthera*. In using the genus name (*Coeloglossum*), Luer reverted to an original nomenclature that first appeared in 1820, and this species is now the only one in the genus. The word, *Coeloglossum*, is Greek for "hollow tongue" and refers to the shape of the lip and its connection to the nectary or spur, which, in this species, is short and rounded. The specific name, *viride*, simply means "green." All parts of this orchid are ordinarily green, but the lip sometimes may be yellowish or, in northern parts of the range, even reddish. The common name that I use is green frog orchid, although others use satyr orchid or long-bracted orchid, referring to the green frog orchid's similarity to the northern variety of the tubercled orchid, *Platanthera flava* var. *herbiola*, which is also called long-bracted orchid.

The flowers of this species are small, rounded, and downward tilted. They are tilted so much that, without getting down to ground level, it is difficult to tell if the flowers are blooming. The lip is narrow and entire, with a cleft in the end. In the southern Appalachians, green frog orchids vary in size from just a few inches to about a foot (approximately 10–30 centimeters) tall. Plants stand up straight and usually have several leaves. A long bract extends beyond each flower.

Green frog orchids have several varieties. One, var. *viride*, is circumboreal, meaning that it ranges around the world in the high latitudes of the northern hemisphere. The variety that exists in the southern mountains, (var. *virescens*, which means "green"), extends across Canada and southward into the Rocky Mountains and into the Appalachians, where it is, however, quite scarce. Var. *virescens* is distinguished by long floral bracts, whereas the northern var. *viride* has short bracts.

A small cluster of green frog orchid plants grows in southwest Virginia.

Since the green frog orchid has no showy colors to catch the eye, it is often unnoticed. My wife and I have found several sites in southwest Virginia. At one of those sites, we first spotted just one plant. But when we got out of the car to look, we found 43 plants of various sizes scattered about the road bank.

The range of this orchid is sporadic in the South. For instance, in Virginia, green frog orchid is found in the northern part of the state, but then the range seems to skip the Roanoke River Valley. Locations reappear in the middle to upper elevations of the New River Valley and areas in the headwaters of the Holston River. Green frog orchid is known from three counties in West Virginia, four counties in North Carolina, and only one county (Carter) in extreme northeastern Tennessee. The plant has not been found in Kentucky.

Look for green frog orchid plants to start blooming in

The flowers of green frog orchid face downward, making it difficult to tell if the plants are in bloom.

mid-April, while some plants will still be fresh into May. In the southern mountains, this orchid is most often found near hemlock and in a damp but fairly open area bordering northern hardwoods at medium elevations.

Corallorhiza (Haller) Chatelain

The coralroot orchids

The scientific name for this genus is derived from the resemblance of the rhizomes to knotty, branched coral. *Corallorhiza* is the combination of two Greek words meaning "coral" and "root" (rhizome), thus "coralroot" is a literal translation. There is no real root in the coralroots but a chunk of rhizome material (actually an underground stem) massed with a fungal partner.

Coralroot orchids have no leaves to carry on photosynthesis. However, according to Dr. John Freudenstein of Ohio State University, coralroot orchids have been found in laboratory analyses to have minute amounts of chlorophyll. Northern coralroot orchid, *C. trifida*, is green and does have some chlorophyll in the stem and flowers. But coralroot orchids get practically all of their nourishment from underground. This is accomplished through a relationship between the rhizome and a fungus partner. The fungus helps break down energy from other sources and then transfers this energy to the orchid. Formerly, such plants were said to be saprophytic, or living off dead and decaying organic material in the soil. But more recent studies have shown that there are no flowering plants that are completely saprophytic.

Coralroot orchid plants are slender, with the height varying among species up to about 12 inches (approximately 30 centimeters). Flowers stand horizontally away from the stem on a short pedicel and wither quickly after pollination. The ovaries then swell and droop on the stems. Each species is known to have a yellow form, often referred to as an albino. In all but one species, the yellow form is entirely yellow or yellow-green and the lip is pure white. In the striped coralroot orchid, *C. striata* Lindley (not found in the southern Appalachians), the yellow form has a pale yellow lip. Typical flowers of most coralroot orchid species in the United States and Canada are beautiful. Their small size sometimes hides the loveliness within, but a close look will reveal a delicate beauty of earthy colors accented with a pure, snowy white lip most often spotted with enchanting magenta markings.

With the discovery of Bentley's coralroot orchid in 1996, there are now seven species of coralroot orchids known across Canada and the United States north of the subtropical zones. Five of these species are found in the southern Appalachian Mountains.

Corallorhiza bentleyi
John V. Freudenstein

Bentley's coralroot orchid

Bentley's coralroot orchid is a new species that was first discovered by the author in 1996. It was formally named and described in 1999 in the botanical journal *Novon* by Dr. John Freudenstein of Ohio State University. Type specimens are deposited in the Ames Herbarium at Harvard University, which houses one of the most complete collections of orchid specimens in the world, and in the Massey Herbarium at Virginia Tech in Blacksburg, Virginia. This new species is thus far known from only one

Bentley's coralroot orchid is the most rare orchid species in the southern Appalachian Mountains.

location in the world and is further distinguished by being the only known species of orchid in the United States or Canada with entirely cleistogamous, or nonopening, flowers.

The closed flowers found in Bentley's coralroot orchid indicate a well-developed system of self-pollination. The stem and the ovaries of each flower are a rich chestnut or mahogany color. The plain sepals and lateral petals are usually the same color but can be tan, giving the flowers a "two-tone" aspect that is quite unusual in the genus. Out of 60 plants studied at the site over the four seasons since the discovery of this orchid, five plants have had all-yellow flower parts. Forcing open a typical flower of Bentley's coralroot orchid reveals a lip color of a pure, immaculate, or unspotted, melon yellow. This color is similar to the lip color in the yellow form of the striped coralroot orchid, *C. striata* Lindley, but is not seen anywhere else in

The flowers of Bentley's coralroot orchid are cleistogamous, or closed.

the genus in the United States or Canada. According to Freudenstein, the orchid most similar to Bentley's coralroot orchid is a small variety of the striped coralroot orchid that grows in Mexico and is called *C. striata* var. *involuta* (Greenman) Freudenstein. This Mexican variety, however, has flowers that open normally.

Plants of Bentley's coralroot orchid are approximately 6–8 inches (approximately 15–20 centimeters) high, about the same height as autumn coralroot orchid, *C. odontorhiza*. But Bentley's coralroot orchid plants present a heftier or "beefier" appearance. Stems and pedicels are thicker and ovaries longer than in autumn coralroot orchid. The flowers of coralroot orchids often go unnoticed because of their small size, but most species have exceptionally beautiful flowers. Still, because the flowers do not open, the outward appearance of Bentley's coralroot orchid does not rank with the obvious beauty seen in other

This flower of Bentley's coralroot orchid has been artificially opened, exposing the melon yellow lip.

species of the genus. The relative plainness of the flowers is easily explained by the fact of their self-pollination. Flowers that do not need pollinators do not need to expend energy trying to be attractive.

Plants of Bentley's coralroot orchid have been found to bloom from the middle of July to early August. This flowering time corresponds closely with the blooming time for spotted coralroot orchid, *C. maculata*, found in the same vicinity. But morphologically, or in its physical appearance, Bentley's coralroot orchid does not resemble the spotted coralroot orchid at all. Spotted coralroot orchid flowers have a white, spotted, three-lobed lip, while the lip of Bentley's coralroot orchid is entire, with no spotting. Spotted coralroot orchids have a very definitive physiological structure called a mentum—a small, spurlike ridge on the bottom at the base of the flower. Bentley's coralroot orchid flowers have no mentum.

Dr. John Freudenstein studies Bentley's coralroot orchid at the only known site.

In West Virginia, where the single site for Bentley's coralroot orchid is located, the environment is very acidic. Plants grow along a well-shaded trail among a thin cover of grass. The underlying rock is sandstone, and there are open, eroded pockets of pure sand scattered about the area near the site. Nearby is a veritable wilderness of rhododendron thickets, called "hells" in the southern mountains. Hardwood trees in the area consist mainly of white oak and chestnut oak. Strangely, not a single pine tree of any kind is found anywhere near the site. Only four plants were found in the original discovery of Bentley's coralroot orchid in 1996, but that number has increased each year, and as of July 1999 there were 27 plants present at the site.

Without question, Bentley's coralroot orchid is the most rare orchid species in the southern Appalachians. Its discovery has enhanced our knowledge and extended the variety of our native orchid flora. John Freudenstein's assistance in determining the status of this new orchid species following its discovery was invaluable and made an important contribution to our appreciation of the orchids of the southern mountains.

Corallorhiza maculata (Rafinesque) Rafinesque

Spotted coralroot orchid

Spotted coralroot orchid's specific name, *maculata*, means "spotted" and refers to the spots on the lip and other flower parts. The tiny flowers are exquisitely beautiful. In the more typical form, spreading lateral petals, along with like-colored sepals, curve about the upward-arching column as if to guard its treasury of pollen at the tip. In the southern mountains, the flowers and the stem are most often a straw or golden color but many are bronzy red. In other areas across North America, flowers can be brownish or even deep red.

The three-lobed lip of spotted coralroot orchid is distinctly different in shape from the other members of the genus. The two side lobes appear as small teeth on either side of the lip. The margins of the much larger middle lobe are parallel to one another, and the forward edge of the lip is scalloped, with a fine lacy texture. But beckoning irresistibly to any admirer is the fabulous snow white color of the lip, which is further showcased by several vivid red-purple spots of varying shape, size, and position. No two flowers—even on the same plant—have an identical pattern of lip spots. With bright sun, especially backlighted, spotted coralroot orchid flowers "seem to glow from an internal light source," according to William Petrie in his *Guide to Orchids of North America*. Though it usually takes some knee-bending effort to fully appreciate spotted coralroot orchid flowers, they are definitely worth a close-up look.

The plants of spotted coralroot orchid are slender and vary widely in height from about 8 inches to sometimes over a foot (approximately 20–30 centimeters). The flowers stand out horizontally from the stem on a slightly arching pedicel. The dried plant, with its opened capsules intact, often overwinters and is sometimes present when the following year's flowers open.

The continental range of this orchid in its typical variety is widespread: the northeastern United States, western Canada all the way east to the Maritimes, south into Cal-

Several tiny but beautiful flowers grace the leafless stem of the spotted coralroot orchid.

ifornia in the west coastal mountains, the Rockies, and the Appalachians. Other varieties of spotted coralroot orchid are generally more localized. In the southern mountains this orchid is fairly common west of the Blue Ridge escarpment except that it becomes scarce in northeast Tennessee and is not known in Kentucky.

The habitat for spotted coralroot orchid is usually in rather dry, mature acid woods, often with accompanying members of the heath family like rhododendron and mountain laurel. The typical variety blooms about mid-July.

Though spotted coralroot orchid is found all across

Below, left:
Corallorhiza maculata *(Raf.) Rafinesque, spotted coralroot orchid*

Below, right:
Corallorhiza maculata *(Raf.) Rafinesque var.* occidentalis *(Lindley) Ames, early spotted coralroot orchid*

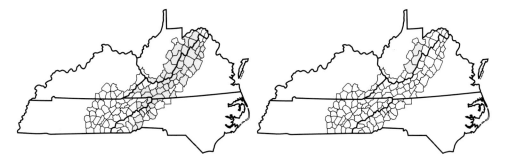

The lip of the typical spotted coralroot orchid flower has straight, parallel margins and articulated side lobes.

North America and is relatively common in the southern Appalachians, the plant is not well known. This anonymity is attributable to several factors, in particular its small size, its natural camouflage, and its preference for dark woods where few summer visitors venture.

Many varieties of spotted coralroot orchid have been described across the continent, some more formally recognized than others. In the southern Appalachian Mountains, in addition to the typical spotted coralroot orchid, one variety and one other form have been recognized.

The scientific name for early spotted coralroot, var. *occidentalis* (Lindley) Ames, means "western," but it is also found in the East. According to Dr. John Freudenstein, this variety is morphologically different from the typical spotted coralroot orchid and is clearly differentiated by the shape of the lip, which has spreading, less parallel sides and a more rounded appearance. Another distinc-

Above:
The yellow form of spotted coralroot orchid is known from only one site in the southern Appalachians.

Left:
Found rarely in the southern mountains, the early variety of spotted coralroot orchid features a rounded lip on each flower.

tion is that this variety blooms much earlier (late May to mid-June) and is recognized in the southern Appalachians from only three counties in West Virginia: Greenbrier, Randolph, and Pocahontas, where I have seen it along the boardwalk at the Cranberry Glades.

Yellow, or albino, spotted coralroot, forma *flavida*, is more often found toward the area of the Great Lakes and in California. Its occurrence in the southern Appalachians is extremely limited, only one very small population located in Virginia. The entire plant is either a bright yellow or yellow-green with lovely flowers that have an unspotted, or immaculate, snow white lip. This yellow form is similar to, and sometimes confused with, the northern coralroot orchid.

Autumn coralroot orchid flowers can be either cleistogamous (left) or chasmogamous (right).

Corallorhiza odontorhiza (Willdenow) Poiret

Autumn or fall coralroot orchid

The specific name, *odontorhiza*, combines two Greek words meaning "toothed rhizome." Some plants demonstrate just that when exhumed, a rounded toothlike projection extending from the normal rhizome. The common name is taken from the season in which this orchid blooms, about mid-August through October in the southern Appalachians.

Autumn coralroot orchid flowers are particularly unusual because there are two distinct types. The term "cleistogamous" is applied to the type with flowers that do not open themselves for pollinators and thus self-pollinate. The type that does open is termed "chasmogamous," which is the normal open condition of any mature flower. It is peculiar to this orchid species in the southern mountains to have both types of flowers.

Autumn coralroot orchid plants are typically 6–8 inches (approximately 15–20 centimeters) high and have

Small, inconspicuous plants typify the autumn coralroot orchid.

very thin stems. Stems and ovaries are often a greenish color, indicating the presence of some chlorophyll. Pedicels and ovaries are narrow, which makes these plants well camouflaged in the grasses and woods of late summer and early fall. The flowers of autumn coralroot orchid are very small and, except for the white lip, are usually a drab purplish-green in color. Open flowers somewhat resemble the shape of those of a tiny Wister's coralroot orchid, *C. wisteriana*. The lip, when observable, is small and very white, with pale purple blotches. But plants with flowers that have immaculate white lips are sometimes found. The fertilized ovaries quickly cause the flowers to droop on their pedicel. Although sometimes present in a particular area in large numbers, these orchids are usually inconspicuous.

Autumn coralroot orchid has an extensive range across the eastern half of the United States, but its numbers de-

The so-called albino, or yellow, form of autumn coralroot orchid is rare throughout the range of the species.

crease in upper New England and around the Great Lakes. This species does not extend into the southernmost part of Florida. It is a common plant in most of the southern mountains, but notably few locations are known in extreme southeastern Tennessee. This orchid is variable in habitat and found in evergreen, mixed, or even all-deciduous, fairly open woodlands. Autumn coralroot orchid also favors mossy banks on roadsides. The plants are more often found in well-drained or even dry soil on ridges with scrub oaks and pines but also may be seen in relatively moist situations.

The yellow form of autumn coralroot orchid, forma *flavida*, is very rare and was first described in 1927 by the well-known botanist Edgar Wherry. This variant form is yellow throughout, except for the lip, which is white and usually unspotted. The few plants of this form that I have observed bloomed in mid-October and were in dry soil with pines and white oak trees. A location for this form in Summers County, West Virginia, was shown to me by Bill Grafton of the University of West Virginia and is the only site I know in the southern Appalachians.

The specific name, *trifida*, refers to the lip, which is three-lobed. But the three-lobed appearance is much less pronounced than in the spotted coralroot orchid, *C. maculata*. In the north country, where this orchid is locally abundant, plants of northern coralroot orchid are a more brownish-yellow color, and the white lip of the flower usually has several pale purple spots. In the southern part of its range, northern coralroot orchid is more green or yellow-green, and the flowers have an unspotted white lip. The plants in the southern Appalachian Mountains belong to var. *verna*, which means "spring," the time of its flowering. This varietal difference was pointed out in the early 1800s by Thomas Nuttall.

Corallorhiza trifida Chatelain var. *verna* Nuttall

Northern coralroot orchid

The height of northern coralroot orchid can vary widely even within the same area. I have seen plants in the Cranberry Glades of West Virginia that were no more than 3 inches (approximately 7 centimeters) high, while, at the same time, on the other side of the glades, I found plants that were about 9 inches (approximately 23 centimeters) tall. The stem and all petals and sepals, except for the white lip, are either green or yellowish-green. This coloring contains some chlorophyll and likely helps the plant to be a bit less dependent on its fungus partner than are other species in the genus.

Like all chasmogamous, or open-flowered, forms of coralroots, the northern coralroot orchid's small flowers are spreading and stand perpendicularly to the stem. This species blooms in late May or early June in the southern mountains and is the first of the "northern" orchids to do so. And even though this orchid blooms early here, its flowering precedes that of the early spotted coralroot orchid, var. *occidentalis*, by only a few days.

Although the continental range of the northern coralroot orchid is extensive—covering the entirety of Canada, Alaska, and the northern United States, and extending into the southern Rocky Mountains—the plant is exceptionally rare in the southern Appalachian Mountains. Northern coralroot orchid, as far as is known, reaches its

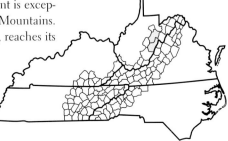

Northern coralroot orchid is common in the North but exceptionally rare in the southern Appalachians.

southern range limit at the Cranberry Glades in Pocahontas County, West Virginia. At the time of the publication of Strasbaugh and Core's revised edition of the *Flora of West Virginia* in 1970, the Cranberry Glades were the only location known for northern coralroot orchid in the southern mountains. But since then, new West Virginia locations have been discovered in Randolph County in the vicinity of Gaudineer Knob. This species remains one of the premier rare orchids of the southern Appalachians.

Although northern coralroot orchid is found in many habitats in the North, it has a specialized habitat in the southern mountains. It is found exclusively in sphagnum bogs or nearby acid, swampy spruce woods.

For several years, a collection from near the Skyline Drive in Albemarle County, Virginia, was mistakenly labeled as *Corallorhiza trifida*. This was a significant listing, being the only one from anywhere in the southern Appa-

Viewed close-up, northern coralroot orchid is seen to have flowers with very small side lobes on each lip—unlike the spotted coralroot orchid flower, which has pronounced side lobes.

lachians outside West Virginia. However, upon closer scrutiny by several botanists, the plants were determined to be the very similar yellow form of spotted coralroot orchid, forma *flavida*. This mistaken listing appeared in the *Atlas of the Virginia Flora* from the time of the first edition in 1977 until the issuance of the third edition in 1992. This is the way botanical records can sometimes get scrambled and is a strong argument in favor of the constant updating of records. It was a disappointment to lose a rare species from Virginia's floral record, but the replacement of the northern coralroot orchid by the yellow form of spotted coralroot orchid was even better. This made the Albemarle County site the only one for the yellow form of spotted coralroot orchid in the entire southern Appalachian Mountains.

The drab, earthy colors of Wister's coralroot orchid can make the species difficult to locate.

Corallorhiza wisteriana Conrad

Wister's coralroot orchid

This orchid takes its specific name from botanist Charles J. Wister (1782–1865), who first collected the species in 1828. According to Luer, the original collection site within the city of Philadelphia is now "a maze of intersecting concrete speedways."

The flowers of Wister's coralroot orchid are small, somewhat rounded, and do not spread very far open to pollinators. Rather than spreading widely, the petals and sepals face forward, leaving only the lip to protrude. However, the blooms are not closed (cleistogamous) in the fashion of autumn coralroot orchid, *C. odontorhiza*, or Bentley's coralroot orchid, *C. bentleyi*. The coloring in separate populations may vary from maroon to brownish or caramel. The tiny lip is bright white with purple blotching. Some plants have flowers with pure white lips. The margins of the lip are curled or rolled inward.

The plant of Wister's coralroot orchid has a slender

Typical flowers of Wister's coralroot orchid have only partially spreading lateral petals and sepals.

raceme of flowers and is quite variable in height, up to about 12 inches (approximately 30 centimeters). The several flowers stand horizontally on the stem and, typical of the entire genus, wither quickly after pollination. The flowers droop on their pedicels as the ovaries enlarge. Populations of these orchids have a reputation for erratic behavior. The numbers can vary from over a hundred plants in one year to only a handful of individuals the next. Look for Wister's coralroot orchid in rocky areas or on cliffs of limestone or dolomite, often along creek sides that are subject to flash flooding. After one particularly bad flood along Jennings Creek in Botetourt County, Virginia, in November of 1985, populations of Wister's coralroot orchid were not seen again in the area for over a decade.

Wister's coralroot orchid has an interesting range across the continent. It is known from the southeastern United

A rare double-lipped flower is sometimes encountered among plants of Wister's coralroot orchid.

States but then skips to the southern Rockies, where the range extends from western Wyoming to Mexico. I have seen plants of Wister's coralroot orchid in the Grand Tetons of Wyoming, in Rocky Mountain National Park in Colorado, and in the shadow of Mount Rushmore in the Black Hills of South Dakota. In the southern Appalachians, there are several records from eastern Tennessee and from three counties in Kentucky: Letcher, Wayne, and Pulaski. In North Carolina, this species is known from only three mountain counties, Buncombe, Madison, and Surry, and it is infrequent in Virginia's mountains. Although present in a few locations in mid–West Virginia, the plant has not been recorded from the Allegheny front mountains that are within the range of this book.

Wister's coralroot orchid can begin to bloom in late March in the southern mountains and continue well into June in its northernmost locations in the Rocky Moun-

Toler's coralroot orchid, or the yellow form of Wister's coralroot orchid, is exceptionally rare throughout the range of the species.

tains. In its current status, Wister's coralroot orchid must be considered uncommon to rare in the southern Appalachians. Although known locations are widespread, their actual number is relatively few. This scarcity is most likely due in part to the orchid's nonshowy posture, which causes it to be frequently overlooked.

Toler's coralroot orchid, C. wisteriana forma toleri S. Bentley, is a newly named form of Wister's coralroot orchid. Being yellow throughout, except for the bright white lip, this form corresponds to the yellow forms found in other coralroot orchid species. A single population numbering only four plants of this form was discovered in Botetourt County, Virginia, in 1995. The plants were observed growing among a population of typical Wister's coralroot orchid that numbered about 150 individuals. Although the population of typical Wister's coralroot orchid reappeared in the next four seasons, the Toler's coralroot orchid did not. This form is named for G. R. (Bobby) Toler of Roanoke, Virginia, who was the first to recognize the significance of the yellow plants.

In order to formally document *C. wisteriana* forma *toleri* as a newly named orchid form, the following systematic description is provided:
Corallorhiza wisteriana Conrad forma *toleri* S. Bentley, f. nov.

A *C. wisteriana* Conrad differt caulibus sepalis petalis lateralis flavidus et labellis ex purpureo albus.

Type: United States. Virgnina. Botetourt County: Along Jennings Creek, 16 May 1995, S. Bentley, *s. n.* (Holotype: VPI).

Cypripedium Linnaeus

The lady's slippers

Lady's slippers are aptly named. They look like small shoes. The genus name, *Cypripedium*, is a combination of two words, "Cypris," another name for Venus, the goddess of love and beauty, and the Latin word "pedis," meaning "foot." The common name of lady's slipper thus is actually a loose rendering of words that could literally be translated as "Venus's or lady's foot."

All lady's slippers in North America bloom in the spring. Although spring is April and May in the southern Appalachians, it isn't spring until late June in some of the higher elevations in the Rockies, both in the United States and in Canada. In Newfoundland, several species of lady's slippers are still in bloom in mid-July.

There are eleven members of this genus found across the United States and Canada, from Alaska to Newfoundland and from near-tundra situations in the North to sweltering habitats near the Gulf Coast. Lady's slippers are, however, absent from the lower parts of the southeast and Florida. This genus is represented by five species in the southern Appalachian Mountains.

The orchids in this genus are slow-growing plants, sometimes taking three or four years to develop from seed to flowering plant. Lady's slippers, so long as the proper fungal partner and adequate sun and water are present, are highly adaptable as far as soil conditions are concerned. Some species are found in strongly acid situations like bogs or dry, oak-pine ridge sides, yet others do best in more basic conditions. Some lady's slippers live in acid soils in one area but thrive in the more basic soil environs of another.

The pink lady's slipper, *C. acaule*, is one of two species that have strictly single-flowered plants. The majority of the lady's slipper species, however, regularly have more than one flower on the same plant. All but one of the lady's slippers in North America normally have what are referred to as "connivant" lateral sepals. This means that the lateral sepals are fused together as one large sepal (called a synsepal) located underneath the pouch, or slipper. One species found in the Great Lakes region and the Northeast has lateral sepals that are not joined.

Most North American lady's slipper species have several leaves that ascend along the length of the stem. The pink lady's slipper is one of two species that have basal leaves. All the lady's slippers have leaves that are rather strongly ribbed or veined in parallel lines.

The most fascinating characteristic of the lady's slippers is the lower petal. This lower petal, or lip, is formed as an inflated pouch or "slipper." In addition to having this unusual shape, it bears the prominent coloring that makes each species particularly enchanting. Except for the pink lady's slipper, all the lady's slippers of North America have a rounded, comparatively large hole at the top of the slipper. The pink lady's slipper, in contrast, has a thin, slitlike opening in the front of the pouch. The lady's slippers have a small triangular or shieldlike structure called a staminode at the base of the lip. The staminode is actually a non-pollen-bearing stamen and stands in front of the column as if protecting the pollen behind.

Lady's slippers have been favorites in many cultures. In *The Native Orchids of the United States and Canada*, Luer described how children in Mexico, for example, have been known to play with the lady's slippers by picking the flowers and making boats of the inflated pouches.

Cypripedium acaule Aiton

Pink lady's slipper or pink moccasin flower

The specific name for pink lady's slipper literally means "stemless," "a" meaning "without" and "caulis" being the Latin word for stem. Michael Homoya in his *Orchids of Indiana* stated that this name refers to "the flowering stalk and basal leaves attached directly to an underground rhizome, giving the impression that the plant is without a stem." This name has been used since at least the 1800s.

Opposite:
A chorus line of pink lady's slipper plants decorates a North Carolina roadside.

Left:
A vertical slit in the center of the slipper must be separated by the pollinator in order to enter the pouch of the pink lady's slipper.

The pink lady's slipper, often called pink moccasin flower, is among the better known and more widely admired plants in the spring forest. This species is a commonly found treasure on most spring wildflower walks, yet most folks are surprised when they learn that these plants are orchids. The solitary pink to purple inflated lip, or slipper, standing atop the bare stem distinguishes this plant from any other. The slipper has a hidden vertical slit in the front center rather than the circular opening at the top exhibited by other lady's slippers. The pollinator, often a bumblebee, must separate this frontal slit in order to enter the pouch. Once inside, the insect finds that the only exit is through the tiny opening on either side at the top of the slipper. This exit requires the insect to squeeze against the pollen bearing column, and the pollinia are thus attached to the pollinator, ready for a ride to the next plant.

Pink lady's slipper flowers are oriented more vertically along the stem than those of any other species in the genus in North America.

The flower's sepals and lateral petals are spreading and either a darker shade of chocolate brownish-green, maroon, or light green. The flower of pink lady's slipper is more vertically oriented in relation to the stem than are flowers of the other lady's slippers. Sometimes the growth pattern becomes deformed and a plant with two flowers or one flower with two pouches may be seen. Also, on occasion, plants with a white slipper (often referred to as an albino form) are found, but this occurs more commonly in the northern part of the plant's range. On one trip to Maine's Mount Katahdin, where the northern terminus of the Appalachian Trail is located, I found many more white-slippered plants than pink-slippered ones.

The paired basal leaves are strongly pubescent and clearly parallel veined. These basal leaves are also diagnostic for pink lady's slipper since all the other lady's slippers in the southern Appalachians have leaves ascending the stem. The size of pink lady's slipper plants is highly variable, often depending on the plant's exposure and soil situation. In dry, oak-pine woods, I have seen plants no more than 3 inches (approximately 7 centimeters) tall, and in swampy acid woods, I have seen plants whose flowers were as high above the ground as a person's knee. Most plants fit somewhere in between.

In the southern part of the southern Appalachian Mountains, pink lady's slipper comes into bloom by mid-April, and at higher elevations like Mount Rogers in Virginia and Spruce Knob in West Virginia, plants can be found in prime bloom in mid-June.

Sometimes pink lady's slipper can be found in tremendous numbers. The Blue Ridge Parkway from the North Carolina–Virginia border southward for at least fifteen miles has population after population along the roadside in late May.

Pink lady's slipper ranges across the northeastern United States and eastern Canada, including the Maritime Provinces. An interesting arching band of its range extends to the northwest into the Northwest Territories but excludes the Rocky Mountains. In the southern Appalachians, pink lady's slipper probably exists (whether or not it has yet been recorded) in every mountain county covered by this book.

The habitat for pink lady's slipper is highly variable but usually within the bounds of rather acid situations in the southern Appalachians. Dry, oak-pine woods are the usual habitat of choice, but plants can be found around the moist borders of boggy areas as well. At higher elevations, pink lady's slipper often stands exposed in heath and red spruce areas. This plant obviously enjoys strongly acid situations in the South but, in the northern part of its range, it is one of the orchids that does a complete turnaround and is common on more basic soils.

Some native orchid species seem to have an affinity for growing near one another. It is a good bet that when one finds a location for pink lady's slipper, large whorled pogonia, *Isotria verticillata*, is likely close at hand—and vice versa. Pink lady's slipper also often establishes itself close to downy rattlesnake plantain, *Goodyera pubescens*. Not only do certain native orchids have a fondness for one another, but other wildflowers are often closely associated with certain orchids. When you are in territory where you see Clinton's lilies, *Clintonia umbellulata*

(Michaux) Morong; lily of the valley, *Convallaria montana* Rafinesque; early pink azalea, *Rhododendron pericylmenoides* (Michx.) Shinners; or Indian cucumber root, *Medeola virginiana* L., you are in pink lady's slipper territory. A good working knowledge of which plant species enjoy one another's company can be very useful.

The natural profusion of pink lady's slipper in the southern mountains and the ease with which it can be located has led to its exploitation and destruction in certain areas. Some of the better-known mail-order nurseries have in the past offered this orchid for sale. The truth of the matter is that the nurseries were employing mountain folks to dig the orchids, sometimes by the hundreds. Although some of this still goes on, a number of states have now passed laws to protect their native flora from the once-booming business of removing native orchids from the wild.

Cypripedium candidum Muhlenberg ex Willdenow

Small white lady's slipper

According to Carlyle Luer, the specific name for this species, *candidum*, is from the Latin "candidus," which means "shining white." From this same origin, we get the English word "candid," which has several meanings. Among those meanings are "unbiased" or "uninfluenced," as in a free or candid opinion. But candid can also mean "pure" or "unblemished"—as surely the pure white pouch, or slipper, of this small, beautiful orchid is.

The single site known in the southern Appalachian Mountains for the small white lady's slipper was found in southwestern Virginia in 1992. This amazing discovery, made far outside the plant's normal range, parallels the discovery of the eastern prairie fringed orchid, *Platanthera leucophaea*, which also exists in only one site in the southern mountains. These two orchid species coincidentally are found in similar habitat situations in the Midwest.

The flower of this lady's slipper is, of course, dominated by the chaste white coloring of the pouch. The sepals and lateral petals are a yellowish-green etched in brown or maroon. The two lateral petals are each formed in a loose spiral and hang down from either side of the pouch. Regularly there are deep red or purple spots around the opening of the pouch that serve to accent the plain white loveliness. The flower very much resembles its close relative,

The small size and pearl white color of the slipper distinguish the small white lady's slipper.

the yellow lady's slipper, C. *parviflorum*. In fact, the flower of the small white lady's slipper and the small variety of the yellow lady's slipper, var. *parviflorum*, are very nearly the same size. There is also little difference in shape among the flowers of the two species. Luer put forth the probability that the small white lady's slipper evolved as a species when isolated from populations of yellow lady's slipper during the periods of glaciation.

Healthy plants of the small white lady's slipper may reach about a foot (approximately 30 centimeters) in height. Several bright green, heavily ribbed leaves sheathe the stem very closely. There is normally only one flower per plant, but twin-flowered plants are not uncommon in some areas. Plants with more than one flower have the flowers arranged one over the other and aligned in the same direction.

Small white lady's slipper probably enjoys more expo-

sure to the sun than any other lady's slipper in North America. The few sites where it is found in its historical range are in swamps and damp fields or meadows with marl or limestone-based soils. Sites are occasionally found among a few trees, but these are likely declining populations that will remain viable only as long as it takes for the trees to grow to a size that will shade out the orchids.

Because of this orchid's close kinship with the yellow lady's slipper, hybrid plants are known from areas where the two parent species overlap. These areas are typically in the Great Lakes region.

Small white lady's slipper blooms at the same time of year, about mid-May, throughout its entire range, which is so small that large differences in elevation or latitude are not a factor in determining bloom time. The small white lady's slipper ranges across the upper Midwest to as far south as Iowa and Missouri, with a few small populations spread a bit farther to the south. One population in Hardin County, Kentucky, was considered, until just a few years ago, the only site south of the Ohio River. But recent discoveries of this orchid in western Maryland (Washington County) and in southwestern Virginia (Montgomery County) have extended its range.

The habitat of the Maryland site, on a precipitous limestone cliff, and the Virginia site, in a rather barren limestone area, are particularly dissimilar to the traditionally described grassy prairie habitat of small white lady's slipper. This is a likely indication that this orchid species once had a much wider range and possibly enjoyed a much more varied habitat as well.

As people have moved onto the small white lady's slipper's traditional prairie habitat and drained the land for agriculture, this species has paid a tremendous price. In fact, although large populations may be found in some specific areas, this small orchid is listed as threatened or endangered over its entire range. The ancestral prairie habitat is all but gone and so, too, is the small white lady's slipper.

Double-flowered plants are common among populations of Kentucky lady's slipper. Note the light-colored sepals and lateral petals.

This impressive orchid draws its specific name from the area where it was originally discovered, Kentucky. The plants themselves were known to botanists for many decades but considered only a variant of the yellow lady's slipper, *Cypripedium parviflorum*. The early botanist Constantine Samuel Rafinesque apparently saw the plants in the 1800s. He described them once in the 1820s and then again in 1833. There is even an extant photo of the orchid that dates from 1939. A name offered in 1971, *C. daultonii* Soukup, or Daulton's lady's slipper, was for a while used by many botanists, including William Petrie in his *Guide to Orchids of North America*, which was published in 1981. That same year, C. F. Reed described the species and named it *C. kentuckiense*.

Kentucky lady's slipper plants are very large. An individual can be 30 inches (approximately 76 centimeters) tall. The leaves have heavy parallel veining and are often

Cypripedium kentuckiense
C. F. Reed

Kentucky lady's slipper or southern lady's slipper

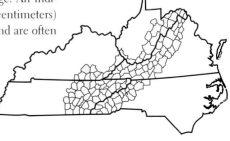

3 inches (approximately 7 centimeters) wide, every bit the size of those of the showy lady's slipper, *C. reginae*. At 4 inches plus (approximately 10 centimeters) measured from tip to tip, the Kentucky lady's slipper flower is quite large, probably the largest orchid flower north of the tropics and certainly the largest in the southern Appalachians. The flowers of the Kentucky lady's slipper may be either single or double, one above the other. The inflated lip is normally a pale or buttery yellow but may sometimes be white. Through years of personal observation, I have noted that the color usually intensifies with the age of the flower. The slipper has several unusual characteristics that add much personality to each flower. The rim of the opening is unevenly ridged, giving one a sense of an uneven mountain horizon. The front of the slipper presents a blunt profile, as if someone had flattened its nose. Inside the slipper, beautifully dark, broken lines of mahogany color run parallel to one another. Lateral petals may be a yellowish green or a dark maroon and spiral several inches down from either side of the slipper.

Kentucky lady's slipper begins to bloom about May 10 in the southern Appalachians, while it may be a bit earlier in its deep southern range of Texas and Louisiana. The range extends up the Mississippi Valley to the latitude of Kentucky and Tennessee, where a limited number of locations are present.

In the southern Appalachians, no sites are known for the Kentucky lady's slipper in West Virginia or North Carolina. All the sites in the area covered by this book are found on the Cumberland Plateau in northern Tennessee and southeastern Kentucky. One remarkable site is known, however, from the area of the western shore of the Chesapeake Bay in Virginia. This extraordinary location was known for some time but, again, the plants were dismissed as a form of yellow lady's slipper until Martha Case of the College of William and Mary began work on the site and published her findings in 1996.

Although it is known from many sites throughout its range, the Kentucky lady's slipper has to be considered rare in the southern mountains. One location in Scott County, Tennessee, normally has 200–300 blooming plants each year, but most other sites have no more than a couple dozen plants.

Kentucky lady's slipper is invariably found in wooded, overgrown floodplains along streams. Of course, this

Opposite:
A cluster of Kentucky lady's slipper plants, each with dark mahogany brown lateral sepals and petals, stands in a Tennessee floodplain.

The late J. T. Baker cross-pollinates a Kentucky lady's slipper at his "Mother Lode" site in 1990.

choice of habitat invites destruction from flash flooding, as I once witnessed in eastern Kentucky. It is a sad sight to see such beautiful orchids bent to the ground, with the slippers filled with mud and water.

The late J. T. Baker of Scott County, Tennessee, was responsible for locating the largest population of Kentucky lady's slipper plants known in the southern Appalachians, which he named the "Mother Lode." The name has become well established among wild orchid enthusiasts of the area, and the site has become well known, having even been photographed, for example, by Baker's cousin, former Tennessee senator and accomplished photographer, Howard Baker.

As far as I am concerned, the Kentucky lady's slipper is the prize orchid of the Cumberland Plateau. The fact that its habitat is in exceptionally remote and rugged country leaves little doubt in my mind that, as interest in our native orchids grows, more sites will be located. Considering the surprise discovery of the new site in eastern Virginia, far, far away from the traditional range, all sorts of possibilities present themselves. Who knows where the next Kentucky lady's slipper plants will be discovered unfurling their spellbinding blossoms.

In the June 1994 issue of the *American Orchid Society Bulletin*, Dr. Charles J. Sheviak published a study that finally separated the North American yellow lady's slipper from its European counterpart. Citing morphological differences, Sheviak proposed the already published (more than 50 years earlier) scientific name *Cypripedium parviflorum* be used for North American plants. The specific name, *parviflorum*, means "small-flowered." This nomenclature has now found wide acceptance and has been used by the majority of orchid authorities in the last few years. The traditional European name for yellow lady's slipper, *C. calceolus*, was the name used by Luer in *The Native Orchids of the United States and Canada* (1975). The specific name, *calceolus*, means "tiny shoe" and had been applied to the yellow lady's slipper in Europe even before Linnaeus published his *Species Plantarum* in 1753. The reference to a little shoe has been the basis for another common name for this orchid, yellow moccasin flower.

In the field, yellow lady's slipper plants are easy to distinguish. The inflated yellow pouch is unmistakable. The waxy-textured surface of the slipper can give the impression of freshly sculpted, fine porcelain. The lateral petals are drawn alongside the pouch, each in a downward curling spiral, as if to adorn the handsome golden yellow slipper with intriguingly elegant laces. Often there is subtle red spotting inside and out to further decorate the slipper.

Yellow lady's slipper plants are usually about 12–15 inches (approximately 30–38 centimeters) high, and a relatively large percentage of individuals will have two flowers adorning the top of the plant. A few specimens in which three flowers are present have been reported. Multiple flowers are placed one over the other and oriented in the same direction. The strongly veined leaves clasp the stem as they ascend and spread diagonally.

In Canada, there are places where yellow lady's slipper

Cypripedium parviflorum Salisbury (syn. *C. calceolus* Linnaeus)

Yellow lady's slipper

Opposite:
A group of yellow lady's slipper plants gathers around an old stump.

Left:
Its bright yellow color distinguishes the yellow lady's slipper. Note the spiraling lateral petals.

plants can be found by the thousands. These northern areas provide singularly limestone habitats, such as those on the Bruce Peninsula of Ontario and the Great Northern Peninsula of Newfoundland. But as one travels south, one finds that the yellow lady's slipper has gradually reversed its habitat and adapted itself to drier, more acidic situations.

The flowering season begins in the lower elevations by the middle of April in places such as along the road between Sugarlands and Cade's Cove in the Great Smoky Mountains. But in the nearby higher northern habitats of the Smokies, it can be June before the yellow lady's slipper plants reveal their beauty. In the Virginia mountains near my home, I begin to look for yellow lady's slipper flowers in May.

Most botanists have, for some time, accepted two primary varietal differences for the yellow lady's slipper. The

A yellow lady's slipper with unusually wide lateral petals grows in a woodlot in southwest Virginia.

large yellow lady's slipper refers to the variety named var. *pubescens* (Willdenow) Knight, which, except for the lip, describes the downy (i.e., pubescent) parts of the plant: the stem, leaves, sepals, and petals. However, both the large and small varieties are likewise pubescent. The name var. *parviflorum* refers to the small yellow lady's slipper, which is generally considered a more northern variety.

The *Atlas of the Virginia Flora*, in its third edition in 1992, did not recognize more than one variety of yellow lady's slipper. West Virginia orchid people have recognized only one location of the small variety in the region covered by this book. That location is in Greenbrier County, which has a great deal of northern habitat. Some plants of the small variety have also been reported by Dennis Horn in the high mountains of western North Carolina. It makes some sense, I suppose, that if the small yellow lady's slipper is a more northern variety, it would need to be distinguished in an area with prominent northern habitat. Because records are incomplete on the varieties of the yellow lady's slipper in the southern Appalachian Mountains, only one map for the species, without regard to variety, is included in this book.

By far, the large yellow lady's slipper is the more common variety in the southern Appalachians. It is found in all parts of the region, although locating a population may prove difficult. Yellow lady's slipper plants are absolutely mesmerizing to me. They are my favorite wildflower. More than any other species, this one is responsible for my fascination with the study of wild orchids. There is something about this lovely yellow orchid that grabs me and pulls a chunk out of my time no matter where I am when I spot it or what other plant I might be pursuing at the moment. All this beauty, especially when encountered by chance, makes for a wonderful experience in the woods.

Cypripedium reginae Walter

Showy lady's slipper or queen lady's slipper

The showy lady's slipper is most appropriately named. The word *reginae* means "queenly" and is derived from the same origin as the word regal. With unmatched beauty, the flower of this orchid is regal indeed.

This orchid will certainly catch one's eye as one comes into its presence. A showy lady's slipper flower can be over 3 inches (approximately 7 centimeters) across, although most are a bit smaller, and is sometimes quaintly tilted to the side, as if the queen is amused at being approached. The petals and sepals are of the purest white, except for the slipper, with its fine texture of smooth velvet, which exhibits the most exquisite pink-purple hues known to abide in nature. Very rarely, flowers are reported that are all-white. The staminode, which deceptively shields the pollen, is a pearly white color with a pale yellow blush and is usually adorned with purple-red blotches. Never is one's first encounter with the showy lady's slipper to be forgotten.

Sometimes individual showy lady's slipper plants will be nearly 30 inches (approximately 76 centimeters) in height. The wide, strongly veined leaves offer a fitting throne upon which the showy queen sits. The stem and leaves are densely pubescent, with fine hairs that may

The unmistakable flowers of showy lady's slipper are among the more beautiful wildflowers in North America. Plants are often double-flowered.

cause skin irritation for some people. Often there are double flowers on one stem and, very rarely, three. Unlike other members of the genus with multiple flowers, showy lady's slipper plants have the flowers arranged side by side rather than one over the other. One can find showy lady's slipper blooming in the southern mountains in early to mid-June, that is, if one can find it at all.

In Minnesota, the showy lady's slipper is the state flower. Across the lower part of eastern Canada and the upper Northeast in the United States, this orchid enjoys a widely varied existence. In some areas of the North, there are often enough individuals in populations for dozens to be clumped together. But in the southern Appalachians, the showy lady's slipper is extremely scarce. Old records do show that there was a wider range years ago than there is today. Although Luer's range map in *The Native Orchids of the United States and Canada* included Ken-

The showy lady's slipper plant is tall, with wide, heavily veined leaves.

tucky, showy lady's slipper has never been confirmed there. There is a location, however, in Claiborne County, Tennessee, which borders the Bluegrass State, so it is not inconceivable that the plant might exist somewhere in the rugged Cumberlands of Kentucky. Tennessee's only other location is on the shore of South Holston Lake in Johnson County, the far northeastern corner of the state.

Only two North Carolina counties (Macon and Jackson) have locations for showy lady's slipper. And there are likewise only two West Virginia counties (Tucker and Greenbrier) that have voucher specimens to document populations. Happily, my home state of Virginia seems to have been more fortunate. There are old records for three counties in northern Virginia, all either extirpated now or nearly so. But very exciting is the discovery of two new sites for this queenly orchid, both in southwest Virginia. In the late 1980s, my friend Doug Ogle found a location

The author photographs a population of showy lady's slipper.

in Washington County, a site now protected by the Nature Conservancy. This site is in the Tennessee River drainage area of Virginia. A bit further east in Virginia's mountains, and in the New River drainage area, a new site was discovered by the author in 1993. For me, the personal discovery of a location for showy lady's slipper was the fulfillment of a decades-long dream. This site is within the George Washington and Jefferson National Forests. Unfortunately, it is right by the roadside and exceedingly vulnerable. But with these new discoveries comes much hope for still more sites to be found in the future.

Typical of the more populated northern habitats, both new Virginia sites for showy lady's slipper are associated with arborvitae or white cedar swamps. But a well-known site in Greenbrier County, West Virginia, is in a rather flat, poorly drained area of a mixed hardwood forest, another example of this orchid's persistence and adaptability. Showy lady's slipper is a beautiful and most welcome addition to the native flora of the southern Appalachian Mountains. The mere presence of such an exceptional plant commands some measure of reverence for the natural world around us.

According to Carlyle Luer, the scientific name for this genus comes from ancient traditional names used in Europe, the historical home of plants in this group. Luer stated that *Epipactis* is the name that was given to a plant used in Greece "perhaps to curdle milk" several hundred years before the birth of Christ.

Two members of the genus *Epipactis* are found in North America. The species we have in the southern Appalachians is a European invader that was introduced in the region around 1879. The second member of the genus is a native American species found along the West Coast and in wet areas of the drier regions of the southwestern United States.

Epipactis Swartz

The helleborine orchids

The specific name, *helleborine*, comes directly from a European genus *Helleborus*, which is in the buttercup family, Ranunculaceae. The *Helleborus* genus is widely represented in this country by the imported ornamental plant called Lenten rose, *Helleborus orientalis*. Helleborine orchid has been used for medicinal purposes, just as have some species in the *Helleborus* genus. In *The Native Orchids of the United States and Canada*, Luer stated: "This common orchid, which recalls the hellebores more by its ancient uses than by its appearance, has been familiar to Eurasians for centuries. Concoctions from its roots and rhizomes have been used as remedies for various ills, including gout."

The flowers of helleborine orchid are about 1 inch (approximately 2 centimeters) across and are not conspicuous. The coloring is usually a dull green but is sometimes yellowish with blushes of pink or purple. The lip is "pinched" in the middle, creating a semi-inflated pouch that is spread more widely open than the pouch of the lady's slippers.

The helleborine orchid plant is slender but has a stout stem that may be 2 feet (approximately 60 centimeters) tall. Leaves of this species are dark green, lanceolate, and strongly parallel veined. There may be several dozen flowers in the raceme that covers the top third of the plant. The flowers hang downward and are arranged on one side of the stem. Lower flowers in the raceme often show signs of withering well before the upper buds open.

Helleborine orchid blooms in the southern Appala-

Epipactis helleborine (L.) Crantz

Helleborine orchid

Helleborine orchid is not native to North America but has established itself in several areas of the continent, including the southern mountains.

chians in June. This orchid is highly adaptable to varying soil conditions. Its favored habitat is shady woods, but it also likes the edges of streams and ponds. Helleborine orchid is sometimes seen in the wild in unlikely places, such as waste areas and dumps, and has also been known to show up in flower beds for which soil has been brought in from another area.

Helleborine orchid has become so common in the northeastern United States and eastern Canada that it is considered a weed in some circles. But there are still relatively few documented sites for helleborine orchid outside cultivation in the southern mountains. The number of locations is growing, however. One location is listed for the eastern Tennessee mountains, in Sevier County near the Great Smoky Mountains. There have been no listings for the Cumberland Plateau of Tennessee or the Cumberland Mountains of Kentucky. Research by friends of mine from Pennsylvania who are writing a book on orchids has turned up no records for the Allegheny ridges of West Virginia, and only one site is listed in the entire state. Helleborine orchid is not even mentioned among the flora of North Carolina. As late as 1992 only three counties were listed for this orchid in Virginia.

Galearis Rafinesque

Galearis spectabilis (L.) Rafinesque

Showy orchis

The genus name for this orchid, *Galearis*, is derived from a Latin word meaning "helmeted," in reference to the sepals and lateral petals that come together to form a hood or helmet over the opening to the nectary. The specific name needs little explanation: *spectabilis* means "showy" and has the same origin as "spectacular," which, indeed, is fitting when applied to showy orchis.

Wildflower guides and orchid books, up until the publication of Luer's *Native Orchids of the United States and Canada* in 1975, used the scientific name *Orchis spectabilis* L. for this species. Luer chose to return to the name *Galearis*, first proposed by Constantine Samuel Rafinesque around the beginning of the nineteenth century. Showy orchis is the only representative of the genus in North America. However, the genus does contain one other species, which is found in east Asia.

The bloom time for showy orchis comes late enough in the spring that the leaves and blossoms of many other plants are already on the forest floor. Showy orchis begins to bloom in the lower part of the southern Appalachians in early April, and flowering is usually complete across the entire region by the first of June. Its small size and shy coloring can mask the unusual beauty of showy orchis and make the plant a bit more difficult to locate. Learn to recognize the leaves, which sometimes appear weeks ahead of the blooms.

The small flowers of showy orchis are exceptionally attractive. The helmet that "protects" the pollen and nectary is a velvety purple, sometimes a bit pinkish. The lip is elongated, about an inch (approximately 2 centimeters) in length and, in most plants, of the purest white. Often the whiteness of the lip appears as a waxy, translucent, pearly white. Either way, the royal purple helmet is perfectly complemented by the white lip. In various plants, some degree of the purple color will trickle onto the lip, either a purple line or purple edging. And sometimes the

A showy orchis plant, with seed capsules from the previous season, grows in a West Virginia forest.

lip may be all-purple, which makes for a most interesting variation in an already striking flower. Rarely, all-white flowers are encountered. Each blossom of the raceme of flowers has a bract that extends above the bloom and often remains recognizable well into the fall, when the flowers have long withered away.

The structure of the showy orchis flower presents a bit of a challenge to pollinators. The lip is reduced in size as it extends backward and becomes a thin spur in which nectar is stored at the very tip. Since few pollinators can reach the entire length of the spur, they have adopted the interesting technique of biting a tiny hole in the end of the spur. I have noticed the same type of hole in the top of flowers of Dutchman's breeches, *Dicentra cucullaria* (L.) Bernh., and squirrel corn, *D. canadensis* (Goldie) Walpers. I once took a flower from a plant of showy orchis and bit off the very tip of the spur. Immediately I experi-

The beauty and easy availability of showy orchis make it a favorite wildflower all across its range.

enced a wave of overly sweet nectar, reminding me of the taste I had experienced from wild honeysuckle when I was a boy. What a source of pure joy this must be to a beetle.

The short rhizome of showy orchis normally sends up a pair of basal, rounded or nearly oval leaves with a single stem in between (sometimes a third leaf may be present). The stem, in healthy plants, may exceed 8 inches (approximately 20 centimeters) in height, but most stems are shorter. Flowers are few in number, seldom exceeding six or eight. Showy orchis can form beds of a few dozen or more plants arranged closely together, but usually plants are scattered singly or in very small groups.

The continental range of showy orchis is from the upper Midwest to New England and, very rarely, into southern Canada. The range then extends southward, with the plant becoming more frequent in that direction.

The contrasting colors of the wide white lip and the purple lateral sepals and petals of showy orchis are a charming addition to the early spring woodlands.

However, this orchid is absent from the deep South. In the southern Appalachian Mountains, showy orchis is a rather common orchid. It probably exists in every county covered by the scope of this book, and indeed it has already been reported in the vast majority of them. This commonness does not in the least detract from its beauty or appeal, though. Showy orchis remains a favorite of every longtime wildflower lover and is a featured plant at almost all the spring wildflower pilgrimages held across the region each year.

Look for showy orchis in open deciduous woods where rich soil is available. The requirements for this orchid species generally include a more basic soil than many of the orchids of the southern mountains will tolerate, although I have seen healthy populations in rather acid situations. Showy orchis is particularly fond of woods where poplar trees have a strong standing. Poplar woods are also a favorite spot for morels or, as mountain folk call them, "dry land fish." Many times I have been searching poplar woods for showy orchis and stumbled on a fine supply of tender morel mushrooms. And many times I have gone to poplar woods after a spring rain to look for morels for my wife's spaghetti sauce and come upon a new population of showy orchis. It is difficult to beat this springtime combination.

My favorite site for showy orchis is one I found on my own in Monroe County, West Virginia, where my father grew up. It is along an old abandoned railroad bed. My grandfather worked for the railroad, and no doubt both he and my father had traveled the route many times. I cannot help but wonder if they ever had occasion to get off the train near the spot where Potts Creek divides from Big Stony Creek. If so, perhaps they, too, enjoyed the display of showy orchis that surely lined the roadside years ago.

Goodyera R. Brown in Aiton
The rattlesnake plantains

This genus was named for seventeenth-century English botanist John Goodyer. Four species within it are found in North America. One ranges from the Great Lakes region west across Canada, dipping south into parts of the Rockies and along the West Coast to mid-California. Another species extends from the Great Lakes eastward across the northeastern United States and Canada to Newfoundland. There are two species represented in the southern Appalachian Mountains.

Each species grows a rosette of leaves that, after accumulating enough energy, sends up a flowering spike (a process that could take five or six years). The old rosette then rapidly withers away. But the plant has sent out an underground "shoot" that will result in the beginning of a new rosette, usually very near the old one. The arrangement of the rosette of leaves in each species of this genus is very similar to that of the common plantain, *Plantago*, and thus the description of plantain has been applied to this entire orchid genus. The members of this genus are just as often referred to, however, as the rattlesnake orchids. The leaves of all the species in this genus have a somewhat rounded, diamond shape. There is a network of markings on each leaf that, to some botanists, resembles the network of scales on the head of a rattlesnake.

Downy rattlesnake plantain has a triangular grouping of flowers arranged all around the upper stem.

Goodyera pubescens (Willdenow) R. Brown in Aiton

Downy rattlesnake plantain or downy rattlesnake orchid

The specific name for this orchid, *pubescens*, means "hairy" or "downy." The entire plant is covered with fine hairs, though the petals, including the lip, are much less hairy than other parts. In this species, each leaf is heavily marked with whitish-green veining and has a wide, whitish midvein. The triangular spike of flowers is tightly packed on the top third of the stem. Flowers are small and rounded and attached in a way that reminds me of a group of tiny Christmas ornaments on a tiny tree. Petals are white and have a waxy texture. The lip itself is somewhat like a miniature pouch or sac with the tip shaped into a short, rounded beak or spout.

Downy rattlesnake plantain begins to bloom in the southern Appalachians about mid-July and lasts into August. Tremendous beds of this plant may form, with dozens of plants present. This species is the most prolific orchid species we have in the southern mountains. Al-

Goodyera **118**

The lip of the downy rattlesnake plantain flower is a ball-like pouch with a short blunt tip.

though downy rattlesnake plantain has not yet been recorded in a few counties in Tennessee, it should be safe to assume that it exists in every county of the region covered by this book. The range on the continent is across the eastern United States except for the deep South and upper Maine.

From dry, shaly pine-oak woods to damp woods with hemlock; from low-elevation areas to well up in the mountains; and from limestone valleys to sandstone ridges, the downy rattlesnake plantain can be found in almost every type of habitat that exists in the southern mountains.

The Blue Ridge Parkway has to be the easiest place to see great numbers of downy rattlesnake plantains. At bloom time, seeing this orchid along the parkway requires only a turn of the head toward the roadside. The nearly foot-tall (approximately 30 centimeter) spikes beckon like candles illuminating the edge of the woods.

From Mabry Mill in Floyd County, Virginia, south along the Carroll-Patrick County border toward Groundhog Mountain, downy rattlesnake plantain is especially prolific. Much of this section of the parkway is lush with rhododendron thickets, one of this orchid's favorite habitats. Often downy rattlesnake plantain is accompanied by pink lady's slipper, *Cypripedium acaule*, and large whorled pogonia, *Isotria verticillata*.

Goodyera repens (L.) R. Brown in Aiton var. *ophioides* Fernald

Lesser rattlesnake plantain or lesser rattlesnake orchid

The specific scientific name for this orchid, *repens*, means "crawling." It is a descriptive name taking its origin from the habit of the rhizome, which moves along under the ground in a creeping fashion before springing through the ground to form a new rosette of leaves. To pull one plant from the ground invariably means to displace others that are attached to the same creeping rootlike system.

Plants of lesser rattlesnake plantain in the United States and Canada belong to var. *ophioides*. This designation distinguishes the New World plants, which have heavy reticulation or markings on the leaves, from European plants, which are uniformly green without markings. Rarely, leaves with no markings are found in North America, and I have seen such plants in Alaska and in the Canadian Rockies.

Lesser rattlesnake plantain flowers are similar to but smaller than those of the downy rattlesnake plantain, *G. pubescens*. The lip of the lesser rattlesnake plantain flower is more like a beak, sharply pointed at the tip and curved downward. Typically, the flowers of this species are arranged on one side of the 5–7 inch (approximately 12–18 centimeter) stem.

Typical leaves of the lesser rattlesnake plantain have heavy but sparse markings. No midvein of the sort found in the downy rattlesnake plantain is apparent. Taking into consideration the one-sided flowering stem and the different leaf markings of the lesser rattlesnake plantain, it should be easy to tell one species from the other.

Both species of rattlesnake plantain bloom at the same time in the southern mountains, about mid-July. The lesser rattlesnake plantain differs greatly, however, from the downy rattlesnake plantain in one important trait. It has a huge range, appearing not only all across the northern United States and Canada but also in northern Eu-

The flowers of lesser rattlesnake plantain are arranged in a row on one side of the stem.

rope and Asia. It is indeed a world traveler. The extension of the range southward in the form of sparse locations in the southern Appalachian Mountains is a likely indicator that the plant migrated here when these mountains were part of a much colder climate and has remained to the present in highly specialized habitats.

In the area of the southern Appalachians covered by this book, lesser rattlesnake plantain is known from relatively few locations, but those locations are fairly widespread. A number of sites are scattered in the North Carolina mountains and in Virginia's mountains outside of the far southwestern part of the state. In Tennessee, this plant is found only in areas along the border shared with North Carolina. Strangely, there are more locations known in southeastern West Virginia than in the northeastern part of the state. The plant has not been reported from Kentucky.

Above:
Side by side, the differences in the markings on the leaves in the rosettes of lesser rattlesnake plantain (top) and downy rattlesnake plantain (bottom) are clear.

Right:
Close examination reveals the heavy pubescence of the lesser rattlesnake plantain and the beaklike lip of the flower.

The relative scarcity of lesser rattlesnake plantain in the southern mountains contrasts markedly with the ubiquity of its very common cousin, the downy rattlesnake plantain. At every site I know for lesser rattlesnake plantain, the downy rattlesnake plantain is always close by. But the reverse is not true. This orchid usually inhabits middle-elevation areas of the southern Appalachians but may occasionally be found at lower elevations. In the North, lesser rattlesnake plantain is considered a strict calciphile, but in the southern mountains it is more often found in acid situations under hemlock or pine trees. A population of lesser rattlesnake plantain may have a relatively large number of individual plants, but the plants do not form large clumps as do plants of the downy rattlesnake plantain.

In these plants, six ("hex," as in "hexagonal") raised ridges that resemble a cock's comb ("alectryon," in Greek) extend the length of the lip and logically lead to the generic name of *Hexalectris*. When literally translated, this is also the source of cock's comb, one of the common names often applied to the members of this genus.

Plants in this genus produce no chlorophyll and thus have no leaves or other green parts. Such plants receive their nourishment in much the same way as do the coralroot orchids *Corallorhiza*, drawing nourishment through their branching rhizomes by way of a fungal connection to an outside source. Stems of these orchids are rather fleshy yet stout. Although flowers of all the species in this group are exceptionally beautiful, with natural earth tone colors such as browns, purples, and maroons, they are not conspicuous within their habitat and may go undetected. Some dark brown stems may give the impression of being dead sticks. But the subtle beauty of the inner flower is quite splendid.

There are five species within this genus in the United States. Only one of them is found in the eastern United States. The other four species of *Hexalectris* are all restricted in the United States to the dry, barren Big Bend area of Texas.

Hexalectris Rafinesque

The crested coralroot or cock's comb orchids

The specific name, *spicata*, means "spiked" and describes the arrangement of the flowers on the stem of crested coralroot. Though most often referred to as a coralroot, this species is not in the same genus as the true coralroot orchids. The flowers are considerably larger than those of the true coralroots of the southern mountains. This orchid, in prime situations, may reach 18 inches (approximately 45 centimeters) in height and present a rather stout appearance. When viewed closely, the flowers of this orchid exhibit a gorgeous combination of dark earthy colors that rival the beauty of almost any orchid. Inside the fresh flowers are subtle combinations of chestnut, purple, maroon, and cream blended together most attractively. The network of purple veining on the lip and the parallel dark veins of the sepals and lateral petals are exquisite.

The caramel-colored sepals are recurved on the several loosely grouped flowers that ascend the top third of the

Hexalectris spicata (Walter) Barnhart

Crested coralroot or cock's comb orchid

Crested coalroot plants have no leaves and often appear from a distance to be dead, brown sticks.

spike. The crowning glory of crested coalroot is the lip. Scalloped at the tip and widening toward the base, the lip is yellowish-white and heavily streaked with bright purple. The six raised ridges that extend the length of the lip are of the finest velvety purple texture. The column presents a creamy white, rounded beacon to pollinators, beckoning as would the bright lamp of a lighthouse. To complete all this fascinating display of subtle-hued beauty is the "throat" of the flower, which is a bright orchid pink.

The flowers of crested coalroot begin to open in late July and last through most of August. The continental range of this orchid extends southward from a line extending from the Mid-Atlantic states to Missouri and reaches westward to Arizona and New Mexico. Crested coalroot is absent from the southern parts of Florida. This orchid has to be considered infrequent in the southern Appalachian Mountains, but it may form large colo-

A *close look exposes the remarkable beauty in the earth tone coloring of the crested coralroot flower.*

nies locally. The infrequency with which it is encountered is undoubtedly due in part to its penchant for picking out-of-the-way, difficult-to-access habitats. Cliffsides, steep ridges, and the tops of mountains seem to be some of the crested coralroot's favorite haunts.

In the southern Appalachians, this species has loosely scattered locations in the mountains of Virginia and several sites on the Cumberland Plateau and in northeastern Tennessee. It is fairly well represented in southeastern Kentucky. However, there are only a couple of mountain locations in North Carolina and two sites in West Virginia's mountains, in Grant and Pendleton Counties.

Crested coralroot's chosen habitat in the southern mountains is generally in more basic soils. Many locations are on dry limestone or dolomite cliff situations. It typically chooses areas with open woods. But on exceptional occasions, this orchid is known to grow in strictly acidic situations, even on solid granite like the very top of Stone Mountain in Alleghany and Wilkes Counties, North Carolina.

Isotria Rafinesque

The whorled pogonias

The Greek word "isos" means "equal," as in isosceles triangle. The latter part of the name comes from "treis," which means "three" and is used here in reference to the three equal-length and very obvious sepals of each flower of the large and small whorled pogonias. The word "pogonia" means "bearded" and refers to the bristles on the lip of the flowers.

The word "whorl" refers to a circular pattern of leaves that radiate out from one central point on the stem. In the whorled pogonias, the whorl of leaves radiates from the top of the stem. The leaf arrangement in this genus of orchids is similar in size, color, shape, and height above the ground to that of Indian cucumber root, *Medeola virginiana* L., a member of the lily family. But the stem of Indian cucumber root is thin, hairy, and solid, while whorled pogonia stems are succulent, smooth, and hollow. Nonflowering plants of the whorled pogonias and the Indian cucumber root are so much alike when viewed from above that they are easily mistaken for one another.

Two members of this genus are found in North America, both restricted to the eastern United States. Their ranges are almost identical.

Isotria medeoloides (Pursh) Rafinesque

Small whorled pogonia

The specific name, *medeoloides*, refers directly to the plant's resemblance to Indian cucumber root, *Medeola virginiana* L. The small whorled pogonia is well known in botanical circles as one of the more rare plants in the eastern United States and is included on everybody's endangered species list.

A sometimes double but usually single flower stands in the center of the small whorled pogonia's umbrella of leaves. The color of the flower's petals and sepals is exactly the same as that of the stem and leaves, a soft, apple green or bluish-green. In contrast, the large whorled pogonia, *I. verticillata*, has flowers that are a very different color from the leaves. The spreading sepals are proportionately shorter, wider, and less pointed than the sepals

The small whorled pogonia is one of the more rare orchids in eastern North America.

of the large whorled pogonia. The three-lobed lip of the small whorled pogonia is greenish-white, with the side lobes forming the tubelike flower. The protruding center lobe is very white and slightly downcurved at the tip. Green veining and a lengthwise yellowish-green crest distinctly mark the surface of the lip. In comparison to the large whorled pogonia, whose leaves point upward, the five leaves of a freshly flowering plant of small whorled pogonia point downward. As the flower ages, the leaves will rise to the characteristic umbrellalike whorl found in young, nonflowering plants.

The hollow, fleshy green stem of small whorled pogonia is usually no more than 8 inches (approximately 20 centimeters) high at bloom time, but I have seen some plants that were less than 4 inches (approximately 10 centimeters) high, including the upright flower. This small size makes hunting the plant very difficult, even in an

Small whorled pogonia plants are often double-flowered. Note that the leaves on a blooming plant point downward.

area where plants have been observed previously. Usually when one plant is spotted, it is best just to stop, sit down, and take the time to completely survey the area. Plants do continue to grow through fruiting, and a plant in an ideal situation will sometimes approach a foot (approximately 30 centimeters) in height while in capsule. The fertilized flower will eventually point skyward, with the ovary forming an upward-pointing seed pod. In the few known sites in the southern Appalachian Mountains, small whorled pogonia blooms in May, but, at any given site, the flowers usually remain fresh for no more than about ten days.

Generally, locations for the small whorled pogonia are best known in the New England states and become fewer as one moves southward. A well-known location in Williamsburg, Virginia, became a historical specialty for botanists in the early part of the twentieth century. This famous site was visited by Frank Morris and Edward A.

Eames in 1925 and written about in their book *Our Wild Orchids*, published in 1929. As more people have become aware of this orchid, the known number of sites has grown. Locations range from northern Georgia to east central West Virginia. Fannin County, Georgia, has at least six locations, most discovered just in the last few years. Two sites were recorded some time ago in the North Carolina mountains, and two new locations were recently discovered in Cherokee County, giving promise that more sites will be found in this southernmost area of the North Carolina mountains, which, incidentally, is not far from the Georgia locations. There is but one site in Virginia's mountains, that being in Lee County, the far southwest corner of the state. And that site was discovered only in 1994. One site from Signal Mountain near Chattanooga represents the only record from Tennessee. Small whorled pogonia still has not been reported from Kentucky. I was fortunate enough in 1996 to be part of a small group of friends accompanying Clete Smith of Pittsburgh when he discovered the first plants of small whorled pogonia ever seen in West Virginia, in Greenbrier County.

Perhaps even more rare than the orchids are the few people fortunate enough to have seen the small whorled pogonia. Its small size, natural camouflage, and rarity all serve to make this orchid one of the most elusive plants in the forest. As more and more is learned about this species, it is becoming obvious that the habitat descriptions will have to be rewritten. Mostly, the plant occurs in deciduous woods, but it can be found in evergreen forests as well. The typical habitat described from New England sites is along braided streams in mature hardwoods and often associated with witch hazel. But I have visited sites in Georgia and Virginia where no water was visible, and no witch hazel either. It just makes sense that as the plant has adapted to varying climates from New England to Georgia, so too has it adapted to varying habitats.

A cluster of large whorled pogonia plants stands sentinel on a Blue Ridge Parkway roadside in North Carolina.

Isotria verticillata (Muhlenberg ex Willdenow) Rafinesque

Large whorled pogonia

The specific name, *verticillata*, means "whorled" and refers to the umbrellalike pattern of this orchid's leaves. In their specific scientific names, both species of whorled pogonias have direct reference to the arrangement of their leaves.

Large whorled pogonia usually has a single flower that stands on a short pedicel at the center of the whorl of leaves, perpendicular to the stem. On occasion there will be double flowers. The three widely spreading sepals are narrow, taper toward the tip, and have a deep purple-brown color that gives the flower the appearance of a large insect. The greenish-yellow, forward-extending lateral petals, along with the lip, form a tubular flower that is open at the end. The lip is three-lobed. The side lobes have a dark purple line running lengthwise toward the base of the flower. The center lobe protrudes beyond the side lobes and has a white, ruffled "skirt" at the end. Pur-

The leaves of blooming large whorled pogonia plants point upward.

ple veining and a crest of green, hairlike bristles adorn the length of the center lobe.

Large whorled pogonia plants often reach a foot (approximately 30 centimeters) in height. The bare stem is reddish-purple, hollow, and succulent appearing. Nonflowering plants will spread their whorl of five leaves (sometimes six) as they come through the ground. However, in a flowering plant the leaves remain tightly together and point upward as if protecting the delicate flower inside. A close look at a budding plant will reveal the already-spreading sepal tips protruding above the leaves. After the flowering process is over, the plant will continue to grow, and the leaves will spread into the characteristic umbrella shape. The insectlike appearance of the flower and the natural coloring of the entire plant make this species difficult to locate.

If spring comes early, the large whorled pogonia will

131 *Isotria*

put in an appearance before the end of April. But in most areas of the southern Appalachians, including my own southwest Virginia area, this orchid blooms in early May. Particularly in this orchid, when the flowers are past their prime, all the parts may remain identifiable for a while but will begin to contort themselves. Sepals will curve in odd directions; the lateral petals will curl back, revealing the inner part of the flower; and the entire flower will often begin to turn so that it faces straight up toward the sky. I consider this species one of the better camouflaged native orchids of the southern Appalachians.

Large whorled pogonia ranges all across the United States east of the Mississippi River, with the exception of the coastal areas of the deep South. In the southern mountains, it is rather common in spite of its reputation for being difficult to locate. There may be dozens of plants in a small area but only a very small number in bloom in a given season. Large whorled pogonia can be found in virtually every mountain county of Virginia and eastern West Virginia. Locations are scattered on the Cumberland Plateau and in northeast Tennessee. All but one of the counties in Kentucky covered by this book have sites. Oddly, relatively few locations are listed from southeast Tennessee or the mountains of North Carolina, although the best location I know is in Alleghany County, North Carolina. It seems to me that the scarcity of known locations in North Carolina's mountains is probably a result of oversight.

The typical habitat for large whorled pogonia is dry, rather acid oak-pine woods. It is often found in association with heaths such as rhododendron and, especially, mountain laurel, *Kalmia latifolia* L. This orchid seems to favor the same habitat enjoyed by the pink lady's slipper, *Cypripedium acaule*. In every site I know for large whorled pogonia, the plants are keeping company with at least a few pink lady's slipper plants. Unfortunately, the reverse is not true.

Liparis
L. C. Richard

The twayblade orchids

The generic name for these orchids is taken from "liparos," meaning "greasy" or "fat," the same word from which we get the words "lipid" and "liposuction." In this instance, the name *Liparis* refers to the shining or wet appearance of the leaves among members of the genus.

This genus is closely related to the genus *Malaxis*, the adder's mouth orchids.

According to Carlyle Luer, *Liparis* has some 250 species around the world. Several species are found in Mexico and the tropics, but only two occur in Canada and the United States north of Florida. Both of these species are represented in the southern Appalachian Mountains. A hybrid between these two species is named Jones's twayblade, *L.* ×*jonesii* S. Bentley, and was discovered in the mountains of North Carolina by the author in the early 1980s.

The two *Liparis* species in the southern mountains have two basal leaves that seem to clasp the stem just above the ground, unlike the true twayblades, members of the genus *Listera*, which have two opposite leaves partway up the stem. The flowers generally extend along the top half of the slender stem in both species. It takes a close look to appreciate the beauty of the very intricate flowers of these orchids.

The specific scientific name for this orchid, *liliifolia*, means "lily-leaved," "folia" coming from the same source from which we get the word "foliage." Its leaves closely resemble the foliage of plants of Canada mayflower, *Maianthemum canadense* Desf., a member of the lily family.

The sepals and lateral petals of the insectlike flowers of lily-leaved twayblade are tiny compared to the broad lip. At first sight, one might think that the lip was the entire flower. Wide at the end and tapering toward the base, the mauve-colored lip is impressive and presents an invitingly broad landing strip for a tiny insect pollinator. Incidentally, the color of the lip inspired one of the older names for this plant, mauve sleekwort. The lip is nearly transparent, sometimes thin enough so that the lateral sepals can be seen through the lip. There is a tiny point in the center of the end of the lip. The beautifully bright purple–colored lateral petals are threadlike, hanging downward on either side of the lip like sadly drooping ears. The light green sepals are widely spread and look as if they are "rolled" lengthwise, giving them a narrow appearance.

The stem of the lily-leaved twayblade plant is usually no more than 6 inches (approximately 15 centimeters)

Liparis liliifolia
(L.) L. C. Richard
ex Lindley

Lily-leaved twayblade

A wide, mauve-colored lip and threadlike lateral petals are characteristic of lily-leaved twayblade flowers.

high, above a pair of somewhat lustrous basal leaves that appear to sheathe the stem. In a large individual plant there may be as many as two dozen flowers. The flower pedicel is horizontal, but the actual flower stands diagonally to the stem. The inflorescence is arranged on the plant in a somewhat triangular shape, slightly tapering toward the top.

Lily-leaved twayblade begins to bloom in the southern Appalachians in mid-May in some areas, but most plants flower more toward the first of June. It may take several weeks to complete the flowering of all buds, from bottom to top, on an individual plant.

The continental range for lily-leaved twayblade is a relatively localized affair. The orchid is found from the Midwest across to the East Coast and down into the upper South. In regard to its range, this orchid appears to be a southern counterpart of its close cousin, Loesel's tway-

A *small cluster of lily-leaved twayblade plants clings to a Virginia mountainside.*

blade, *L. loeselii*, which is generally a more northern species.

In the southern Appalachians, lily-leaved twayblade is rather common, found throughout the area covered by this book. But because of its natural earth tone coloring and small size, this orchid is well hidden within its environs. I feel certain that further discoveries will show it to be even more prevalent than is presently recognized. It is found in virtually every mountain county in Virginia and is widespread in North Carolina. It is found in northeast Tennessee and in the southeastern mountains of Kentucky. Only in the West Virginia mountains are sites sparse and local, but that is likely because the inconspicuous plants have not been discovered.

Lily-leaved twayblade inhabits open, mixed woods in either acid or basic situations. Plants are frequently found along roadsides, usually in well-drained locations. But occasionally this orchid is also found near streams and even in wet meadowlike habitats. Sometimes lily-leaved twayblade is found in the same habitat with the normally water-loving Loesel's twayblade. One location near a well-known site for showy lady's slipper, *Cypripedium reginae*, in Greenbrier County, West Virginia, has both of these species of "fat" twayblades growing side by side in the spongy, wet grasses. There seems to be an exception to any rule where orchid behavior is concerned.

Liparis loeselii (L.) L. C. Richard

Loesel's twayblade or fen orchid

Linnaeus himself named this orchid for Johann Loesel, who was a German botanist in the early 1600s. Loesel wrote a flora of the state of Prussia (which is part of modern Germany).

All parts of this orchid species, including the flowers, are green or yellowish-green. The sepals are spreading and appear to be rolled lengthwise. Lateral petals are drooping and threadlike. The downcurved lip is curled into a "U" shape toward the base but is wider and more flat at the tip. Flowers of Loesel's twayblade are often tilted in different directions.

Plants of Loesel's twayblade are usually no more than 6 inches (approximately 15 centimeters) high and have a narrow cylindrical raceme of flowers above a pair of basal leaves. Living up to the generic name, Loesel's twayblade leaves sometimes appear to have been buttered or smeared with mayonnaise and then wiped clean. Plants are much more easily located in the autumn, when the seed capsules take on a bright yellow color, than in June, when the blooming plants blend in so well with their green, grassy environment.

In the southern Appalachians, Loesel's twayblade blooms in the early part of June in lower-elevation areas. It can bloom as late as the first part of July, however, at higher elevations, such as along the Blue Ridge Parkway near Mount Mitchell in North Carolina and in areas near the Cranberry Glades in West Virginia.

This orchid ranges across the upper Midwest and southern Canada and into the northeastern United States and the lower Maritime Provinces. In the southern mountains, Loesel's twayblade is often unnoticed. Although still not recorded in Kentucky, this species has been found widely in West Virginia's mountains and in five

Above:
The lip of a Loesel's twayblade flower is narrow and bends downward. Note the rolled appearance of the lateral sepals.

Left:
Loesel's twayblade leaves have the glossy or greasy look characteristic of the genus.

mountain counties of northwestern North Carolina. It is sparsely scattered in eastern Tennessee. In Virginia, Loesel's twayblade is known from three northern counties and from five mountain counties in the southwestern part of the state. A survey taken in a normally restricted government defense facility in my home county of Pulaski in 1997 revealed a population of Loesel's twayblade, a new county record.

Loesel's twayblade has to be considered rare in the southern Appalachians. But sometimes large colonies numbering well over a hundred plants are found. For miles along Route 150 in West Virginia, Loesel's twayblade grows in ditches alongside the road. But in two sites in Giles County, Virginia, never more than a handful of plants are seen in any one year. Experience has shown me that the more this species is sought, the more it is found.

The common name fen orchid that is sometimes applied to Loesel's twayblade comes from the orchid's tendency to grow in this sort of habitat, fens being wet, grassy areas well open to the sun. In the southern Appalachians, these conditions can be found along seeps on banksides, especially those that are mossy and in roadside ditches where water stands for at least much of the season. Very infrequently, this orchid does appear well away from any water.

Liparis ×*jonesii*
S. Bentley
Jones's twayblade

This hybrid between lily-leaved twayblade, *L. liliifolia*, and Loesel's twayblade was first discovered by the author in the early 1980s but was not named until 1995.

Jones's twayblade is distinguished by the color and consistency of its petals. The color intensity of the lower petal, or lip, varies, but the color is brown or mauve and usually darker than that of the lily-leaved twayblade. Some plants have lips with an opaque consistency, like those of Loesel's twayblade plants, while others have lips that are transparent enough to reveal the lateral sepals beneath, a feature seen in the lily-leaved twayblade. Beneath each lip on the flowers of Jones's twayblade are obvious veining lines much darker in color than the lip itself. This suggests the same type of veining found in flowers of lily-leaved twayblade. The lateral petals of Jones's twayblade are turned downward, resemble threads, and vary in color from plant to plant from bright pink or purple, as found in the lateral petals of lily-leaved twayblade, to green, which is seen in Loesel's twayblade.

Jones's twayblade is known from only one location, in Alleghany County, North Carolina. Loesel's twayblade and lily-leaved twayblade are known to share a habitat at numerous sites, just as they do at the Jones's twayblade site. But there is no indication that this hybridization has occurred elsewhere. Jones's twayblade is named for my late friend and colleague, J. I. (Bus) Jones of Chattanooga, Tennessee.

Opposite:
The lip of the flower of the hybrid Jones's twayblade has the shape of Loesel's twayblade and the coloring of lily-leaved twayblade.

In order to formally document *L.* ×*jonesii* as a newly named orchid hybrid, the following systematic description is provided:

Liparis ×*jonesii* S. Bentley, hybrid nov.
Planta inter *L. loeselii* (L.) L. C. Richard et *L. liliifolia*

(L.) L. C. Richard intermedia, floribus *L. loeselii* forma similis sed *L. liliifolia* textura et colore similis.

Type: United States. North Carolina. Alleghany County: Along the Blue Ridge Parkway, 11 June 1992, S. Bentley, *s. n.* (Holotype: VPI).

Listera
R. Brown in Aiton

The true twayblade orchids

This genus is named in honor of Dr. Martin Lister, a highly respected English naturalist of the late seventeenth and early eighteenth century.

Eight species in this genus are distributed widely across the United States and Canada. One species is found in North America only on the Bruce Peninsula of Ontario and is generally considered to be an escapee from Europe. One other species is found along the Gulf Coast and lower Atlantic Coast, with a few disjunct populations near the Great Lakes. Two species are represented in the southern Appalachian Mountains. All other members of this genus are cool-climate plants ranging from Newfoundland to Alaska and southward within the Rockies and along the West Coast.

The plants of this genus typically have two sessile, or stemless, leaves (twayblades) opposite one another about a third of the way up the main stem. All species have flowers that are green or brownish-maroon or both. The flowers on orchids in this genus are small. The variation in size and shape of the flower lip or lower petal in each species sets that species apart from the others. Plants are slender and multiflowered, usually with no more than two dozen or so individual flowers. Some hybridization has been noted in this genus, but none is known in the southern mountains.

Listera cordata
(L.) R. Brown in Aiton

Heart-leaved twayblade

Heart-shaped, or cordate, leaves give this orchid its specific name, *cordata*, as well as its widely used common name, heart-leaved twayblade.

The flowers of this orchid species are tiny, less than a quarter of an inch (approximately 5 millimeters) across. Typically, the sepals and petals are all the same color on an individual plant, either green or reddish-brown. In a particular colony, both color forms may be present. Usually the lip presents a different shade of color, sometimes

Tiny insectlike flowers with a deeply forked lip distinguish heart-leaved twayblade.

more intense, sometimes more subtle, than the other flower parts. The sepals and lateral petals flare outward and often have a dark midvein. Toward its end, the lip separates into two narrow lobes that resemble the tip of the forked tongue of a snake. There is a very tiny hornlike structure on either side of the base of the lip.

Healthy plants of heart-leaved twayblade may reach 6 inches (approximately 15 centimeters) in height, but typical plants are usually smaller. Several insectlike flowers stand out along the top third of the slender stem. Typical of the genus, leaves are paired, opposite, and sessile. The stem is smooth below the leaves but has a few sparse hairs above.

This orchid blooms in the southern Appalachian Mountains during the last week of May and into the first part of June. All the locations known in the southern mountains are in areas of cooler climates; thus an early or

The heart-leaved twayblade normally has two opposite heart-shaped leaves about one-third of the way up the stem.

late spring can have a definite and direct influence on the exact date of flowering.

The heart-leaved twayblade is common throughout most of its wide range in northern areas of both the Eastern and Western Hemispheres. This orchid is particularly common in the evergreen forests of the northern United States and adjoining areas of Canada, but it is exceptionally rare in the southern Appalachians.

As much as any other orchid species in the southern mountains, heart-leaved twayblade is intensely pursued. Knowledge of its existence among these ridges has spurred many amateur and professional botanists to spend much time and energy looking for locations, mostly to no avail. This orchid is reported in one historical record in Avery County, North Carolina, but has not been seen there for many years. The plant has never been found in Kentucky, Tennessee, or Virginia, although many botanists expect

A rare third leaf is sometimes encountered on heart-leaved twayblade plants, such as this one in the northern hardwood zone of West Virginia.

to find it in Virginia one day. At present, West Virginia is the only state where one has a good chance of seeing heart-leaved twayblade in the southern mountains. In Strasbaugh and Core's *Flora of West Virginia*, published in 1970, this orchid was reported only from Randolph County. In the meantime, however, Greenbrier County has been added, and several new sites have been recorded in and around the Cranberry Glades and Cranberry Wilderness of Pocahontas County as well as the Gaudineer Knob area of Randolph County.

When it does occur in the mountains of the South, this orchid is found in mixed northern hardwoods, often called the "beech-maple" zone. Usually there are hemlock and rhododendron nearby to help with the highly acidic environs that heart-leaved twayblade requires. The plants, when they can be located, are either in damp moss near a bog or along a stream in forest floor duff made up of beech, maple, and yellow birch leaves. Both in 1997 and in 1998, while in the process of writing this book, I found new populations in Pocahontas County, West Virginia. The proper habitat certainly exists in many areas of the southern Appalachians for more populations to be found, and I believe that growing interest in our native orchids will lead to the discovery of a number of new sites for heart-leaved twayblade in the coming years.

Listera smallii Wiegand

Small's twayblade, kidney-leaved twayblade, or Appalachian twayblade

Small's twayblade is named in honor of John Kunkel Small, who, along with his associate A. A. Heller, came upon the plant in 1891 in western North Carolina. In 1897 Small named the species *Listera reniformis* after the reniform, or kidney-shaped, leaves. However, Karl M. Wiegand termed this name improper because it had been used previously for another plant. In 1899 Wiegand renamed the orchid *Listera smallii*, citing one of Small and Heller's collections from near Blowing Rock, North Carolina. Small was a well-known and influential botanist who wrote voluminously about the southern mountains. His works include *Flora of the Southeastern United States* (1903), *Manual of the Southeastern Flora* (1933), and *Ferns of the Southeastern United States*, which was published in 1938, the year of his death. Small's twayblade is also referred to as Appalachian twayblade in keeping with its general range restriction. This southern Appalachian endemic orchid, as much as any other, is a cornerstone in defining the orchids of the southern mountains.

The flowers of Small's twayblade are made inconspicuous by their small size and subdued earth colors. Plants may be either green or reddish-brown, each usually having like-colored flowers. Individual populations ordinarily are of one color. Occasionally both colors can be found on some blossoms, in which case, the flower presents a rather understated beauty all its own. Each flower stands out from the stem on a short pedicel, or flower stalk, and is typically oriented diagonally to the stem. Sepals are usually reflexed, while the narrow lateral petals spread widely. The lateral petals sometimes exhibit a dark lengthwise midvein. On either side of the base of the lip is a small, rounded, glandlike structure. The lip broadens into two rounded, flat lobes toward the tip and has a tiny "tooth" in the middle of the notch between the lobes.

The wide lobes and small "tooth" in the center of the lip distinguish the flowers of Small's twayblade.

The flowers, as in many orchid species worldwide, superficially appear insectlike.

The thin stem of Small's twayblade may reach 8 inches (approximately 20 centimeters) in height. The leaves are in the shape of a kidney (reniform), giving rise to another common name, kidney-leaved twayblade. The leaves are dark green and have a slightly rough texture. The stem is rather smooth below the leaves but downy above. Flowers are arranged loosely along the upper third of the stem, and the inflorescence tapers somewhat in width toward the top of the plant.

Small's twayblade begins to bloom in some areas of the southern Appalachian Mountains toward the end of June and can still be found fresh at higher elevations near the end of July. One location on Droop Mountain in Pocahontas County, West Virginia, is in prime bloom about the middle of July.

Small's twayblade plants may be either green, like this one, or reddish-brown.

Small's twayblade has a rather limited range across the East. It reaches its northern range limit in mid-Pennsylvania and its southern limit in northern Georgia. The plant is uncommon in the mountains of eastern West Virginia. In Virginia several sites are sprinkled across the northern counties, but then, coming south, the plant virtually skips the Roanoke River Valley only to become rather common again over the remainder of the southwestern part of the state. A small number of sites are found near the border between Tennessee and North Carolina. There are many sites on the Blue Ridge Parkway, especially just north of Mount Mitchell in North Carolina. Small's twayblade still has not been reported from Kentucky.

It is not particularly difficult to locate a population of Small's twayblade when looking in the proper habitat. This orchid is found in strongly acid situations. In lower elevations, it is most often found near streams and growing under rhododendron, *Rhododendron catawbiense* Michaux or *R. maximum* L., and/or mountain laurel, *Kalmia latifolia* L. This species is also found on the high ridges of the southern mountains and, incidentally, nowhere near any water. Small's twayblade is also present sometimes in sphagnum bog situations such as those around the Cranberry Glades in West Virginia.

Small's twayblade, though small, is a much larger plant, with much larger flowers, than the heart-leaved twayblade, *L. cordata*.

Malaxis is the Greek word for "weak" or "soft"; as the name of this genus of orchids it refers to the tender consistency of the leaves of the plants. The common name adder's mouth is derived from the appearance of the flowers. In several species, the reflexed sepals and petals present the lip and pollen-bearing column in a forward position more advantageous for receiving pollinators, a position like that of a snake's tongue.

There are eight universally recognized species within this genus in the United States and Canada, although some authorities now list a ninth species as well. Four of these are found only in drier parts of the country, like the Big Bend area of Texas and along the border that Arizona and New Mexico share with Mexico. One species is found across the northern United States and Canada all the way to Newfoundland. One strictly southern species grows only in the coastal plain in damp, well-shaded, swampy locations from the Mid-Atlantic states to Florida. Another species found only in bogs or in near-tundra situations of the far North is one of the most rare orchid species in all of North America. In the southern Appalachians, we have traditionally had only one species, but some botanists have now recognized a former variety as a new species.

Malaxis plants characteristically are very small, with tiny green flowers. One species in the far southwestern United States has maroon flowers. Some species along the border with Mexico have very tightly clustered spikes of tiny flowers, while eastern and northeastern species usually have more loosely arranged spikes.

Malaxis Swartz
The adder's mouth orchids

Green adder's mouth orchid has the smallest orchid flowers in the southern Appalachians.

Malaxis unifolia Michaux

Green adder's mouth orchid

The origin of the specific scientific name for this orchid is easily discerned. The prefix "uni" ("one") has been combined with "folia" ("foliage") in reference to the fact that green adder's mouth orchid plants have a single, somewhat oval leaf. The leaf base sheathes the stem about a third of the way up.

The flowers of this orchid are very tiny, the smallest of all the orchid flowers in the southern Appalachians. All parts of the flower, including the lip, are green or yellowish-green. The lateral petals are very narrow and reflexed behind the lip toward the ovary. The spreading sepals are linear and about three or four times wider than the lateral petals. The size and shape of the lip can vary from plant to plant and even among individual flowers on the same plant. The two-lobed lip is proportionately wide compared to other parts of the flower. The base of the lip surrounds the column as would auricles or "ears," and the

A single leaf sheaths the stem of the green adder's mouth orchid.

end of the lip is notched with a very tiny "tooth" in the center between the two lobes.

While most plants are about 6 inches (approximately 15 centimeters) high, healthy mature specimens of the green adder's mouth orchid may reach 8–10 inches (approximately 20–25 centimeters). A fresh plant is packed tightly with multiple buds arranged in a flat, spiraling fashion at the top of the stem. The buds open from the outer edge of this spiral as the plant matures upward. The flower arches downward on a comparatively long but very narrow pedicel. This process causes the inner buds to be the last to open and eventually to end up positioned at the top of the plant. Often, ripening capsules will appear in the lower part of the plant well before all of the buds have flowered. Individual flowers can still appear intact even when attached to a ripening capsule.

Some botanists have separated green adder's mouth or-

chid into two distinct species. *Malaxis bayardii* Fernald, Bayard's adder's mouth orchid, is said to have a more cylindrical inflorescence, with the flowers attached to the stem by a very short pedicel. According to Tom Wieboldt, assistant curator at the Massey Herbarium at Virginia Tech, many plant collections taken in southwestern Virginia and labeled as green adder's mouth orchid have been found to have characteristics pointing more toward Bayard's adder's mouth orchid.

In the southern mountains, green adder's mouth orchid blooms appear at lower elevations about mid-June, but in higher areas, such as the grassy balds on the shoulders of Mount Rogers in Virginia, the flowers open in the latter part of July.

This orchid species ranges across the eastern half of the United States and southern Canada to the Maritime Provinces and southward as far as central Florida. This wide range includes all of the southern Appalachians, yet relatively few locations for green adder's mouth orchid are known. The small size of the plant and its tiny flowers make green adder's mouth orchid easy to overlook. Additionally, it blooms in the hot summertime when most folks have given up outdoor pursuits except those connected with a cool, shady spot.

Look for green adder's mouth orchid in open areas of relatively short grass with acidic-dominated vegetation. The place where I have found it easiest to see and photograph this orchid is at Massie Gap in Grayson-Highlands State Park in Grayson County, Virginia. Most sites are in rather dry areas. There are also many sites along the margins of mixed woods, with heaths such as mountain laurel or rhododendron nearby.

Platanthera L. C. Richard

The fringed orchids

The word *Platanthera* is derived from Greek and means "wide anther." The anther is the pollen-producing part of a flower, and the anthers of members of this genus are particularly large compared to those of some of the other genera in the orchid family.

Until the publication of Carlyle Luer's *Native Orchids of the United States and Canada* in 1975, the name *Habenaria* had been used exclusively in this country to designate this group of orchids. But according to Luer, a different arrangement of the pollinia within the flower separates the several species found across the United States and Canada from the true *Habenaria* of more tropical regions. Luer therefore reverted to use of the name *Platanthera*, which had been applied to these orchids in Europe as early as 1818.

Orchids within this genus are first distinguished by a very slender, tubelike elongation of the base of the lip that turns into a spur extending behind the flower. This spur becomes the nectary, and many species are pollinated by butterflies and moths, which have a long proboscis that can probe the length of the spur. As the proboscis probes the spur for nectar, the head or back of the moth rubs against the column, and the pollinia attach themselves in preparation for a ride to another flower or another plant. Plants of this genus are referred to as the fringed orchids because many members have an obvious fine cutting or fringe around the margin of the lip on each flower. But also included are several orchid species whose flowers are unfringed. Some plants have lips which are lobed while others are entire. Most plants in the genus have leaves that are lanceolate (lance-shaped) and ascend the stem. But one species, the pad-leaf orchid, *Platanthera orbiculata*, has large saucer-shaped leaves that lie flat on the ground. All species in this group have their flowers arranged in a raceme or grouping of individual flowers occupying the top third or so of the stem. The number of flowers on an individual plant may vary from a few in some species to well over a hundred in others. Flowers open in order from the bottom to the top of the raceme as the plant's flowering period progresses. The primary time of blooming is mid-summer.

Not including species from the subtropical South, there are approximately two dozen species from this genus in Canada and the United States, giving it the largest number of representatives in nontropical North America. In the southern Appalachians, thirteen separate species are present, many among the loveliest of all our native wildflowers. The flowers of purple fringed orchids, of which there are two separate species, have no rival when it comes to delicate beauty. Almost all members of this genus are brightly colored and are conspicuous within their open field habitat. It is a treat indeed on a warm summer day to come upon a field alive with the glowing colors of these orchids.

The specific name for this species is taken from the Greek words "blepharon," meaning "eyelash" and "glotta," meaning "tongue." The fringed lip of this species resembles an eyelash.

White fringed orchid has two varieties, separated primarily by range. The northern variety, var. *blephariglottis*, is found in bogs and wet meadows eastward from the Great Lakes along the Canadian border and through New England into the Maritime Provinces of Canada as far north as Newfoundland, where it is found in coastal bogs. Plants of this variety are smaller and have a smaller flowering raceme than the southern variety. Named for its conspicuous display in the wild, var. *conspicua*, the southern variety, is found primarily along the coastal plain in savannahs and wet, swampy situations from New Jersey to Texas.

In the South, white fringed orchid is an impressive plant. Healthy plants may be 18 or more inches (approximately 45 centimeters) tall, with a magnificent raceme of white flowers encompassing the top third of the green stem. This unmistakable wand of bright white is visible from some distance away.

Each part of the flower of white fringed orchid is pearly white except for the light yellow–colored pollinia on either side of the column. The lateral sepals are reflexed, which serves to project the pollen-bearing column forward. The lateral petals are very slender and have a slight fringe on the end. Closely aligned with the lateral petals, the dorsal sepal creates a "roof" or covering over the column. The unlobed lip is slender, about an inch (approximately 2 centimeters) long, and heavily lacerated or fringed around the complete margin. But the most striking structure in the flower is the very long, downward-curving spur. The spur is twice the length of the flower itself and very closely resembles the spur length found in the white fringeless or monkey face orchid, *P. integrilabia*.

The white fringed orchid is part of the flora of the southern Appalachians in only two counties in North Carolina, Henderson and Haywood, and one area in Virginia, a group of vast, wet fields called Big Meadows just off the Skyline Drive in Shenandoah National Park in Page County.

Platanthera blephariglottis (Willdenow) Lindley var. *conspicua* (Nash) Luer

White fringed orchid

Opposite:
The white fringed orchid is rare in the southern mountains.

Opposite:
No two yellow fringed orchid flowers have the exact same fringe pattern.

Left:
Yellow fringed orchid is fairly common in the southern mountains in early August.

The specific name for this particular species, *ciliaris*, means "fringed." Although several of the species within this genus have fringe about the margin of the lip and some even on the lateral petals, the yellow fringed orchid is the only one given this literal name.

All parts of the flower of the yellow fringed orchid are a bright yellow or orange color. The species is sometimes referred to as orange fringed orchid. The lateral sepals are reflexed, leaving the dorsal sepal to form a hood or cover over the column. The lateral petals are narrow and project forward to the edge of either side of the dorsal sepal. The ends of the lateral petals are also finely fringed. The unlobed lip is deeply fringed along the entire margin and has an elongated triangle shape, tapering to a point at the tip. No two flowers—even on the same plant—have exactly the same fringe pattern. A long, narrow tube or

Platanthera ciliaris (L.) Lindley

Yellow fringed orchid

spur reaches backward in each flower as an extension of the nectary. (Compare the flower description for the crested fringed orchid, *P. cristata*, below.)

Most healthy plants of yellow fringed orchid are about 15 inches (approximately 38 centimeters) high, but in an ideal habitat this orchid may reach 20 or more inches (approximately 51 centimeters). The flowering raceme itself may reach 8–10 inches (approximately 20–25 centimeters) in length, with dozens of flowers standing out horizontally from the stem. The raceme is slightly triangular, with a rounded appearance on the sides, and becomes more narrow at the top. Buds are tightly clustered about the stem. The flowers open from the bottom up and then arch away from the stem on a narrow pedicel. Several lanceolate leaves ascend the stem and are reduced to bracts upon reaching the flowers.

By late July and early August, yellow fringed orchid begins to make its appearance in meadows and along roadsides throughout the southern Appalachian Mountains. Oddly enough, plants growing in the sweltering heat of the Green Swamp in North Carolina and other areas of the coastal plain give no concession to the high temperatures and bloom at exactly the same time as do plants in the mountains.

This orchid is widespread across the eastern half of the United States and southern Ontario southward to central Florida. In the southern mountains, yellow fringed orchid is relatively common in all parts of the area covered by this book, although some areas may have only small populations. In more than a few areas, yellow fringed orchid populates some meadows by the hundreds, as is the case in one particular wet field in Barbour County, West Virginia. Another small meadow near the Blue Ridge Parkway in Alleghany County, North Carolina, always has more than 50 flowering plants tucked into one tiny corner.

Look for this orchid in open sun but often hiding among weeds of the same height. Though some roadside populations in the mountains appear to be on dry banks, a closer look usually reveals a seep that at least keeps the orchid's roots wet. The plant is not found in open water but does sometimes relish wet, boggy situations. I have found populations of yellow fringed orchid to be particularly variable and vulnerable. One site I know has as many as 50 blooming plants one year and only 3 or 4 the next. Altering the water supply or changing the amount of sunlight can have a definite effect on any native orchid population and, prolific as they are, yellow fringed orchid plants seem particularly susceptible to even small disruptions in habitat.

Platanthera clavellata (Michaux) Luer

Club-spur orchid or woodland orchid

The specific name, *clavellata*, is taken from the Latin word "clavellatus," meaning "club-shaped" and refers to the swollen or enlarged tip of the spur of each flower in this species.

Flowers of club-spur orchid are small and rounded. Lateral sepals are green and slightly spread. The lateral petals and dorsal sepal are also green and combine to shape the top of the flower into a nearly hemispherical hood. The lip is either a lighter shade of green or white, and it protrudes beyond the other parts of the flower. The lip is scalloped on the end, giving it a three-lobed appearance. Individual flowers are often oriented in different directions. The nectary is extended backward into a succulent-appearing and moderately sized spur.

The club-spur orchid plant is inconspicuous because of its green color and slender appearance, which blend in with its grassy habitat. Plants can reach about 10 inches (approximately 25 centimeters) in height and have relatively few flowers in comparison to other members of the genus. The flower raceme is short and restricted near the top of the stem. A narrow, shallowly keeled leaf appears just above the ground on the stem and makes it look as if the stem might have arisen from within the base of the leaf.

In the southern mountains, club-spur orchid blooms in the latter part of July, with some allowance for differences in elevation at specific site locations. It blooms about the same time that female Diana fritillary butterflies, *Speyeria diana*, begin to be seen on the thistles, joining the males who have emerged earlier.

The continental range for this orchid is the eastern half of the United States except for Florida and the Gulf Coast. It is found across southern Canada from western Ontario all the way into Newfoundland, where there is also a dwarf variety that is described by some authors.

Above:
Club-spur orchid plants are small and usually display only one narrow leaf partway up the stem.

Right:
The raceme of club-spur orchid bears few flowers, all parts of which are the same green color.

Over the entire area of the southern Appalachians covered by this book, club-spur orchid is a rather common plant within its habitat.

Club-spur orchid can be found in wet areas of open fields or in woods. Among its favorite haunts are grassy wet meadows, roadside seeps, and the margins of ditches, ponds, or lakes. It also is seen in boggy situations. Club-spur orchid seems equally comfortable in acid or more basic environments and is found at all but the highest elevations of the southern mountains.

The crested fringed orchid flower has a short, triangular, deeply fringed lip.

The scientific name for this species, *cristata*, is from the Latin "cristatus," which means "crested." This refers to the fringe around the margins of all three petals. The fringe seems to remind some botanists of the crest or comb on the head of a chicken.

Crested fringed orchid is superficially similar to the yellow fringed orchid, *P. ciliaris*, described above. On close examination, however, there are some striking differences. First is the obvious difference in size. Crested fringed orchid is much smaller, perhaps 10 inches (approximately 25 centimeters) high and has a smaller, more compact raceme of flowers. The raceme is more cylindrically arranged than that of the yellow fringed orchid, which is shaped in an elongated but somewhat rounded triangle.

All parts of the crested fringed orchid's flowers are yellow or orange. The flowers are tiny, probably no more

Platanthera cristata (Michaux) Lindley

Crested fringed orchid

than a fourth the size of the yellow fringed orchid flower. The spur of crested fringed orchid is very short compared to that of other members of this genus. But the best way to differentiate the two, if there is any doubt after considering their size, is to look at the shape and pattern of the lateral petals. In the crested fringed orchid, the lateral petals are rounded, with fringe around the complete margin. In the yellow fringed orchid, lateral petals are slender and have only slight fringe across the very tip.

It is August before the crested fringed orchid begins to blossom in the southern Appalachians. It is normally a coastal plain species found in the southeastern United States outside of southern Florida. Because some unique geographical situations similar to coastal plain habitat can be found in the southern mountains, this orchid can also be found there, making a special appearance in five counties on the Cumberland Plateau of Tennessee. One location was recorded long ago from Henderson County, North Carolina, and one record exists from Rockbridge County, Virginia. Nowhere else in the southern mountains is this special little orchid to be found.

The typical habitat for the crested fringed orchid in the mountains of the South is in wet meadowlike areas. One site in Van Buren County, Tennessee, is in a small clearing for a power line just off a highway.

I first encountered the crested fringed orchid on one of my many trips to the Green Swamp area of Brunswick County, North Carolina, where it grows in more typical savannah habitat along with many other coastal orchid species. I have the late Bus Jones of Chattanooga, Tennessee, to thank for making me aware of its presence in the southern Appalachians.

Platanthera flava (L.) Lindley
Tubercled orchid

The specific name, *flava*, means "yellow" and refers to the plant's color. One common name, southern rein orchid, is often used because of the superficial similarity of this species to the northern rein orchid, *P. hyperborea* (L.) Lindley. But the term "rein orchid" is inappropriate since neither of these species belongs to the same group as the true rein orchids of the tropics. The name tubercled orchid is better and causes the least confusion.

Plants of tubercled orchid are slender, with the narrow raceme of flowers loosely placed along the top third of the

Above:
The flower of the northern variety of the tubercled orchid displays the typical "bump," or tubercle, at the center of the base of the lip. Note the long bracts that extend well beyond the flower.

Left:
The flower of the southern variety of the tubercled orchid has yellow petals and short flower bracts.

stem. This species may reach a height of 10–12 inches (approximately 25–30 centimeters) and is well camouflaged in tall grasses. I have more than once taken photographs of a population only to find that in the resulting photo the plants were barely discernible.

This species has two distinct varieties that are recognized by most botanists. They are told apart by subtle differences in the lip, the length of the floral bracts, their blooming time, their range, and their habitat.

Below, left:
Platanthera flava (L.) Lindley var. flava, southern tubercled orchid

Below, right:
Platanthera flava (L.) Lindley var. herbiola (R. Brown) Luer, northern tubercled orchid

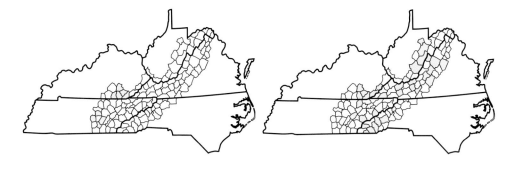

Platanthera flava
var. *flava*
Southern tubercled orchid

The varietal name for this orchid, *P. flava* var. *flava*, literally means "yellow yellow." Petals of the flowers of this more southern variety of tubercled orchid are yellow-green, certainly more yellow than those of the northern variety. In opposition to its more northern counterpart, the southern variety could be called short-bracted orchid, in that the floral bract reaches only about midway of the ovary. The lip is more nearly the same length of the lateral petals, making it look as if all three petals are completely alike. Flowers of the southern variety also have the tubercle at the center of the lip.

The southern variety of the tubercled orchid opens its flowers in late July. This later flowering time has probably evolved in part because its chosen habitat is out of the sun. The southern variety prefers plenty of shade in wet woods or swamps and the heavily shaded margins of wet meadows. It is often found in cover so dense and dark as to make photography quite difficult.

The southern variety of tubercled orchid is found in the southern Appalachians in only a few places in southeastern Kentucky and on the Cumberland Plateau in Tennessee. Coincidentally, both of these areas are regions where the ranges of the two varieties overlap. Other sites outside the mountains are generally located on the coastal plain of the Gulf states except for southernmost Florida. This makes the southern variety of the tubercled orchid a very rare plant in the southern Appalachians.

Platanthera flava
var. *herbiola*
(R. Brown) Luer
Northern tubercled orchid

The more northern plants of this species are designated var. *herbiola* (R. Brown) Luer, which simply means "grasslike," coming from the same source from which we get "herb" or "herbage."

All parts of the small flowers of the northern variety of tubercled orchid are green. The three sepals are about the same length as the lateral petals but more spreading. There is no yellow coloring at all. All three petals are the same shade (usually a lighter green than the sepals), and the slightly recurved lip is noticeably longer than the lateral petals. The lip is three-lobed, but the side lobes are small and tucked away near the base, which gives the impression that the lip is entire. There is an obvious tubercle or "bump" in the center, near the base of the lip. Some botanists have speculated that the tubercle helps

The northern variety of the tubercled orchid hides itself well in its green, grassy habitat.

guide pollinators to either side of the column, where the pollinia are located. The tubercle has, of course, given rise to the most commonly used name for this species. Another common name, long-bracted orchid, is derived from the unusually long bract that extends beyond each flower. The long bract is a reliable field character that will help separate this northern variety of the tubercled orchid from the southern variety.

The northern variety of tubercled orchid blooms in the southern Appalachians in the first half of June. The plant ranges across southern Ontario and the northeastern United States into New England and then southward to northern Georgia.

The northern variety of tubercled orchid becomes infrequent in the southern Virginia and West Virginia mountains. It is found only in a few counties of the southwestern corner of North Carolina and appears in only four counties in the eastern Tennessee mountains. Kentucky records include six counties in the southeastern corner of the state. The preferred habitat for this variety of tubercled orchid is in full sun or very limited shade in open wet meadows. Roadside banks where seeping water keeps the soil wet are also favored. Undoubtedly, this variety is often overlooked because of its capability to hide itself within its environment, but it does often occur in large colonies.

Platanthera grandiflora (Bigelow) Lindley
Large purple fringed orchid

The specific name for this orchid is a simple combination of two Latin words, "grandis," meaning "large," and "floris," meaning "flower": large-flowered. For many years, because of its very close resemblance to the small purple fringed orchid, *P. psycodes*, the large purple fringed orchid was listed as a variety of the smaller species. Today, though, all botanical texts list the two plants as entirely separate species.

The lateral sepals in the large purple fringed orchid flower are spreading but not acutely reflexed. The lateral petals arch upward and, along with the dorsal sepal, form an open hood over the column. Fully mature flower racemes of the large purple fringed orchid are usually wider at the bottom than at the top. The comparatively large and round opening to the nectary is a reliable field character in the identification of this species in the southern Appalachians. (Compare the flower description for small purple fringed orchid below.)

Over the years, I have observed that one of the principal pollinators of the large purple fringed orchid is a common, medium-sized butterfly called the silver-spotted skipper, *Epargyreus clarus*. Many times I have watched while this lively insect busily worked a group of orchids. It is both interesting and amusing to see the butterfly's head with a half dozen or more pollinia stuck fast, protruding like so many horns on some grotesque beast.

Flowering racemes of the large purple fringed orchid are a splendid deep lilac color. But the shade of the purple may vary from plant to plant, and even white-flowered plants have been reported. The lateral petals are widely spreading and very shallowly fringed. The lip spreads openly as a three-lobed, full "beard," which is deeply fringed about the outer margins of each lobe. A relatively

The large round opening to the nectary in the flower of the large purple fringed orchid distinguishes the species.

long spur extends back from the opening to the nectary. The large purple fringed orchid presents itself as one of the most strikingly beautiful flowering plants of the southern Appalachian Mountains. It is always a treat to see this orchid in the wild.

Healthy and ideally situated large purple fringed orchid plants, such as some of those along Route 39 at the Greenbrier-Pocahontas county line in West Virginia, can be well over 30 inches (approximately 75 centimeters) tall and have close to a hundred individual flowers on the top third of the stem.

Large purple fringed orchid begins to flower in the southern mountains about the middle of June in the high country of the Blue Ridge Parkway north of Asheville, North Carolina. A few days earlier, the small purple fringed orchid has already begun to bloom in the same area. Herein lies one of the peculiarities of the purple

fringed orchids. In the northern part of their range—Pennsylvania on up through the Northeast—it is the large purple fringed orchid rather than the small that presents its flowers first each season. No one has as yet offered a good explanation as to why this reversal of blooming time takes place, but this phenomenon is sometimes seen in other plant species as well.

The large purple fringed orchid is found across the northeastern United States, southward through the mountains, and from southern Ontario eastward through the Maritime Provinces of Canada. In the southern mountains, intermittently placed populations occur from middle to higher elevations. This species is not listed from Kentucky and is listed from only two counties in Tennessee, both (Carter and Johnson) in the far northeastern corner of the state. In North Carolina, this orchid is present in several mountain locations, among them the Craggy Gardens area of the Blue Ridge Parkway. Eight counties in West Virginia's mountains have populations of this orchid. The range extends to only a few sites in northern Virginia; then, moving southward, it oddly skips the Roanoke River Valley, and the plant's next appearance southwestward in Virginia is in Giles County, in the Mountain Lake–White Rocks–Cascades area of the New River Valley. Even though large purple fringed orchid has to be considered very uncommon in the southern Appalachians, fairly large localized populations are sometimes found.

Large purple fringed orchid is often seen in somewhat more shaded conditions than are most other members of this genus. Rather partial to more acidic environments, it often frequents shaded wet meadows and seeps that are located near rhododendron. It is also fond of sphagnum bogs. Some plants can be found each season within the bogs of the Cranberry Glades and Canaan Valley areas of West Virginia. Like many other orchids, large purple fringed orchid is particularly sensitive to changes in its habitat. Many plants grow along roadsides and in ditches, which makes them vulnerable to "diggers" and the regular maintenance objectives of the highway department.

This species hybridizes with ragged fringed orchid, *P. lacera*, giving rise to *Platanthera ×keenanii* (see below).

Opposite:
Large purple fringed orchid is one of the most beautiful wildflowers in North America.

The raceme of the yellow fringeless orchid is very small yet may have dozens of flowers.

Platanthera integra (Nuttall) Gray ex Beck

Yellow fringeless orchid

The lip of the yellow fringeless orchid is entire, a fact that accounts for the derivation of its specific name from the Latin word "integer," meaning "entire" or "undivided." This character is something it shares with the white fringeless orchid but that sets it apart from other members of this genus, which have either a fringed lip or a divided lip or both.

The small flowers on any individual plant of this orchid species are all one color, a bright tangerine yellow. The lateral sepals are spreading while the dorsal sepal, along with the nearly oval-shaped lateral petals, serves as a hood protecting the column and the minute opening to the nectary. The path to the nectary opening is protected on either side by comparatively large pollinia. The spur extends about the same length as the ovary. The lip is very slightly serrated around its outer edge, but not fringed.

Examined at close range, the yellow fringeless orchid is seen to have lip margins that, while uneven, are not fringed.

Racemes of this orchid are usually cylindrical but sometimes more narrow at the top. The inflorescence is tightly packed, with up to two dozen flowers, and the entire raceme of yellow fringeless orchid can fit between a person's thumb and forefinger when held in an open parallel position. It really is a small plant, and, with the lack of fringe on the lip of the flowers, there should be no mistaking this species. Mature plants are narrow and may reach about 10 inches (approximately 25 centimeters) in height. The plants are similar in size to those of the crested fringed orchid, *P. cristata*.

Yellow fringeless orchid blooms in early to mid-August. In the southern Appalachian Mountains, this orchid is extremely rare. It has never been reported from the mountains of Kentucky, West Virginia, or Virginia. However, there are records for Cherokee and Henderson Counties in North Carolina. In Tennessee it is known in

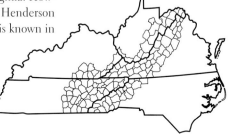

the mountains only from Van Buren, Bledsoe, and Warren Counties, all on the Cumberland Plateau.

The typical geographic range of yellow fringeless orchid is on the coastal plain from New Jersey to Texas. It is an uncommon orchid species in all its range. Only quirks in the geologic makeup of the southern mountains have enabled us to count this orchid among our native flora.

Typical habitat for this species in most of its range is full sun or slight shade in damp savannahs and edges of swamps. In the southern mountains, yellow fringeless orchid has settled for wet grassy meadows. One fabulous site in Van Buren County, Tennessee, is within the cleared margins of a power line. This site is blessed not only with the rare yellow fringeless orchid but also with hundreds of yellow fringed orchid, *P. ciliaris*, and white fringeless orchid, *P. integrilabia*, all growing together.

Platanthera integrilabia (Correll) Luer

White fringeless orchid or monkey face orchid

Like the yellow fringeless orchid, *P. integra*, the white fringeless orchid owes its specific name to the fact that it has an undivided lip. In this case, the Latin words "integer" ("undivided") and "labia" ("lip") are combined to provide the descriptive specific name.

Flowers of white fringeless orchid are large, and the entire flower is snow white. Because of its geographic isolation in the mountains and its similarity to the more prolific coastal species, white fringed orchid, *P. blephariglottis* var. *conspicua*, I would guess that this species evolved from the coastal one. White fringeless orchid has become one of the rare mountain specialties. The lateral sepals are reflexed, but the dorsal sepal and short, narrow lateral petals together form a rooflike structure over the column. The lip is narrow and long, with a slight bend upward about the middle of its length. The primary characters that set this species apart from others are the unfringed lip and an especially long spur, which can be twice the length of the flower. The nectary opening is particularly gaping, like a mouth. The large yellow pollinia provide the "eyes" and the large nectary opening provides the "mouth" that give the flower the appearance that lies behind its other common name, monkey face orchid.

Plants of this orchid may be 15–18 inches (approximately 38–45 centimeters) high but have relatively few flowers. Thus the raceme is short in comparison to that of

Above:
The lip of the white fringeless orchid flower has an upward bend about halfway along its length.

Left:
The raceme of the white fringeless orchid is conspicuous in its habitat. Note the exceptionally long spurs on the flowers.

many other species in the genus. D. S. Correll treated this plant as only a variety of the white fringed orchid in his *Native Orchids of North America*, and Radford, Ahles, and Bell, in *Manual of the Vascular Flora of the Carolinas*, stated that "plants found in the mountains of North Carolina with essentially entire lip are known as var. *integrilabia* Correll." But in *Native Orchids of the United States and Canada* (1975), Carlyle Luer asserted that the white fringeless orchid is as distinct from the white fringed orchid as the purple fringeless orchid, *P. peramoena*, is from the large purple fringed orchid, *P. grandiflora*. Most authorities have accepted Luer's elevation of the plant to the status of species.

The white fringeless orchid is restricted to sites on the Cumberland Plateau of Tennessee and a few other sites in the southwestern part of the Appalachians. Historical records show that this orchid's range was once considerably wider. The small range distribution that exists today naturally makes this orchid a rare plant, although some local populations, such as that on Starr Mountain in southeastern Tennessee, are quite large, with hundreds of plants. The flowering time is about the first week in August. The white fringeless or monkey face orchid is fond of wet meadows, clearings for power lines, and swamps.

Platanthera lacera (Michaux) G. Don in Sweet

Ragged fringed orchid

The Latin word "lacer" means "torn" or "lacerated." It provides the specific name for this orchid species and refers to the deep, irregular cutting of the lip of the flower. The lip's tattered appearance gives the plant its common name of ragged fringed orchid. This species is also referred to as the green fringed orchid.

All parts of the flowers of this orchid are green, but the three-lobed lip is typically a much lighter shade than other parts, sometimes almost white. Each lobe of the lip is fringed or cut into widely varying shapes of hairlike strips. Even on the same plant, no two flowers have the same pattern of lacerations. The longer middle lobe is much less fringed than the side lobes. The lateral sepals are reflexed, while the dorsal sepal and the narrow, unfringed lateral petals form the typical "roof" over the column and nectary. On very rare occasions, plants that have fringing on the lateral petals have been recorded.

Ragged fringed orchid plants are slender and usually about 15–18 inches (approximately 38–46 centimeters) high, with a long, cylinder-shaped raceme. Flower numbers on any one plant can range from about a dozen to several dozen. Plants blend in well with their grassy surroundings, but the lighter coloring of the mature plant often betrays its hiding place.

Plants of ragged fringed orchid may begin to bloom in the southern Appalachian Mountains by the middle of June, with some still fresh in mid-July in higher elevations. The continental range is across the eastern United States except for the deep South. This species also ranges from Ontario across Canada into Nova Scotia and Newfoundland, where large populations are often encountered.

In the eastern part of the southern mountains, ragged fringed orchid sites are fairly abundant, occurring in most all of the mountain counties of North Carolina, Virginia, and West Virginia. But only sparse locations are known

Above:
A typical ragged fringed orchid stands in a midsummer fallow field.

Left:
Close inspection of the ragged fringed orchid flower discloses the pronounced lacerations of the lip.

from southeastern Kentucky, and the plant is infrequent on the Cumberland Plateau and in the mountains of eastern Tennessee.

Ragged fringed orchid prefers open sunshine in fields or wet meadows, though it is sometimes found about the margins of damp woods. In the southern Appalachians, this species is relatively easy to find, but, peculiarly, I have never seen a really large population in any one area.

Look in wet, grassy roadside ditches or open fallow meadows in the southern mountains in late June or early July and you are likely to find ragged fringed orchid.

Ragged fringed orchid, or green fringed orchid, hybridizes with large purple fringed orchid, *P. grandiflora*, and with small purple fringed orchid, *P. psycodes*.

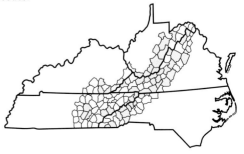

Andrews's fringed orchid is today considered to be the hybrid of the small purple fringed orchid and the ragged fringed orchid. Albert Leroy Andrews, in the first half of the twentieth century, was the first to suggest that there were indeed hybrids between the two species. For many years, any suspected hybrids from these parents fell under this one, all-inclusive name.

Andrews's fringed orchid has many forms and color combinations that reflect characters from both parents, especially as one goes farther north looking at specimens. The most interesting of all the Andrews's fringed orchids I have seen are plants I encountered in Newfoundland and Nova Scotia, which had flowers with purple petals and green sepals. They are beauties indeed.

In the southern Appalachians, plants of Andrews's hybrid are exceptionally rare. This is due, in part, to the fact that the ranges of the parent plants seldom overlap in the South, while in the far North, they overlap commonly. I have only twice encountered the Andrews's fringed orchid in the southern mountains, once in Virginia and once in North Carolina. As is usual with hybrid orchids, one should not expect to find plants in the same number, the same configuration, or the exact same place year after year.

Platanthera ×andrewsii (White in Niles) Luer

Andrews's fringed orchid

Opposite: This lovely specimen, with light green sepals and purple petals, is only one of many color forms of the hybrid Andrews's fringed orchid.

The rare Keenan's fringed orchid is considered to be a hybrid between the large purple fringed orchid and the ragged fringed orchid. The name was not published until 1993. These plants previously were lumped together with hybrids of the small purple fringed orchid under the name Andrews's fringed orchid. This hybrid was first determined in the southern Appalachians by the author, Hal and Helen Horwitz of Richmond, and Bobby Toler of Roanoke, Virginia, in July 1992 in Pocahontas County, West Virginia.

On the occasion of this first encounter in West Virginia, one particular plant fit what would be a classic description of a purple fringed and ragged fringed hybrid: the flowers were exactly the same shape as those of ragged fringed orchid, but they were purple. In this population, quite a bit of variation in flower shape was noted from plant to plant, and sometimes flowers varied widely on

Platanthera ×keenanii P. M. Brown

Keenan's fringed orchid

Right:
The hybrid now known as Keenan's fringed orchid was not formally named until 1993.

Opposite:
This flower of the hybrid Keenan's fringed orchid has the shape of the ragged fringed orchid flower but the coloring of the large purple fringed orchid.

the same plant. Several degrees of hybridization were exhibited, probably attributable to backcrossing. Some of the more interesting flowers were very large, with exceptionally long lateral petals that protruded beyond the upper sepals, giving the flower a "rabbit ear" appearance.

Keenan's fringed orchid was named for Philip E. Keenan, of Dover, New Hampshire, the author of *Wild Orchids Across North America*. New sites for this rare hybrid have recently been found in West Virginia. At this writing, Keenan's fringed orchid has now been studied for eight seasons in the southern Appalachians, and orchidophiles have developed a special interest in this orchid because of its beauty and rarity.

Platanthera leucophaea (Nuttall) Lindley

Eastern prairie fringed orchid

According to Carlyle Luer the specific name for this orchid comes from "the Greek, 'leucon,' meaning white, and 'phaios,' gray or dusky," in reference to the "greenish-white or creamy color of the racemes of flowers." From a distance, racemes of this orchid do appear white, especially in contrast to the greens of its habitat in open meadows. The eastern prairie fringed orchid grows in only one known location in the southern mountains.

All parts of the flowers of eastern prairie fringed orchid are greenish-white, with their coloring lighter on the front sides of the petals and sepals. Typically, the sepals and lateral petals are spread widely. The ends of the lateral petals are rounded on either side, appearing almost two-lobed. The wide, heavily fringed lip is divided into three lobes and spreads downward, openly exposing the large yellowish pollinia. The pollinia are parallel to one another on either side of the opening of the nectary. The flowers broadcast a sweet perfumelike fragrance in the evening hours. Plants in the western part of the range of the prairie fringed orchid were separated in 1986 into a separate species, *P. praeclara* Sheviak and Bowles, and are commonly referred to as western prairie fringed orchid.

Eastern prairie fringed orchid plants can be quite large, often approaching 3 feet (approximately 90 centimeters) tall, with elongated racemes covering the upper third of the stem. The flowers are large and loosely grouped. There are usually several flowers, but typically there are fewer than in most species of this genus.

The normal range for eastern prairie fringed orchid is across the upper Midwest, around the Great Lakes region and rarely into southern Ontario, where it blooms about the last week of June. In the southern Appalachian Mountains, this orchid blooms in mid-June and is one of the region's exceptionally rare orchid species.

Even in its more typical range, eastern prairie fringed orchid has become a rare plant because of human encroachment on the prairie environment. The prescribed natural habitat for this orchid is wet prairies and open marshes. As people find more and more use for former prairie wetlands, fewer areas of this special habitat remain, and unfortunately this is no less true for the eastern prairie fringed orchid than for its western cousin. Historically, wet prairies occurred regularly in Virginia's Shen-

The intricate flower of the eastern prairie fringed orchid is especially striking.

andoah Valley. It is likely, however, that natural evolution of the area's drainage over time and the subsequent massive farming of this wide, fertile land have drained the valley to a point where such prairies are very rare, if they exist at all.

Far from its known range, eastern prairie fringed orchid was discovered in Augusta County, Virginia, in 1980 by Tom Wieboldt, who is now the assistant curator of the Massey Herbarium at Virginia Tech. The site has never been known to have more than a handful of individual plants in bloom in any one season.

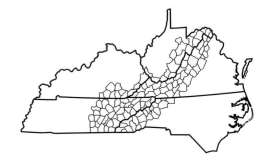

Opposite:
Eastern prairie fringed orchid plants stand out among the tall green grasses of their habitat.

Left:
The lip of the pad-leaf orchid flower is unlobed and unfringed. Note that the viscidium of the pollinarium is visible at the base of the pollinia.

"Orbis" means "circular" in Latin and is the source from which we get the English word "orbit" or simply "orb." It is also the source for *orbiculata*, the specific scientific name of the pad-leaf orchid, inspired by the large, round leaves of this species.

All parts of the flowers of the pad-leaf orchid are a whitish-green, very similar to the color found in the ragged fringed orchid, *P. lacera*; the eastern prairie fringed orchid, *P. leucophaea*; and the club-spur orchid, *P. clavellata*. The lateral sepals are recurved, and the lateral petals are curved upward into a near semicircle around the dorsal sepal. The lip is entire, about three-fourths of an inch to an inch (approximately 2 centimeters) long, and slender. The end of the lip is sometimes curved inward toward the stem. The exposed pollinia are thrust outward as an open invitation to pollinators.

Stems of pad-leaf orchids are usually 12–15 inches (ap-

Platanthera orbiculata (Pursh) Lindley

Pad-leaf orchid or large round-leaved orchid

The smaller variety of the pad-leaf orchid has oval-shaped leaves and flowers that are more widely spaced along the stem than those of the large variety.

proximately 30–38 centimeters) high. The cylindrical raceme of flowers may be tightly packed, sometimes with up to a hundred or so flowers. The paired saucer-shaped leaves are huge compared to other orchid leaves and lie prostrate on the ground. These large leaves are the most distinctive field character of this species.

Many authorities have divided this species into two varieties. According to these botanists, the varietal names are assigned because of several consistent differences found in populations of this species. The large pad-leaf orchid is named var. *macrophylla* (Goldie) Luer, the varietal name meaning "large-leaved." This variety is said to have leaves two times as large as the smaller and has a shorter flower raceme, which is tightly packed with flowers. Most noteworthy is the fact that the flowers of the larger variety have a measurably longer spur on each flower. The smaller variety, var. *orbiculata*, has leaves that

The large variety of pad-leaf orchid has large, round saucer-shaped leaves.

are sometimes more oval-shaped than round but are always smaller. Its flowers are fewer and more sparsely placed along the stem.

The time for blooming of the pad-leaf orchid in the southern Appalachians is from the latter part of June into early July. The species has a rather wide continental distribution, ranging from western Canada (where I happened upon pad-leaf orchid plants on a trail at Mount Robson in British Columbia) across the upper Canadian prairies and into the Maritime Provinces, including Newfoundland, where I have seen the species in Gros Morne National Park. Pad-leaf orchid's range then extends southward through the Appalachian Mountains to Georgia.

Pad-leaf orchid is not well known among wildflower enthusiasts in part, I suppose, because of its usual habit of growing in the deep woods rather than out in the open. Although pad-leaf orchid sites are somewhat widespread, plants are present in limited numbers throughout much of the area covered by this book. It becomes more scarce, however, in the southern reaches of the mountains. It is seldom found in populations numbering more than a couple of dozen plants, and those individuals are typically spaced well apart from one another. Often many more sterile or nonflowering plants (those having leaves only) than flowering ones are present in a population.

Records exist from most of the mountain counties in Virginia. The same is true for West Virginia, except for the counties in the very northeastern corner, which, incidentally, is at lower elevation near the Potomac River drainage. Pad-leaf orchid is present in only two counties of the extreme northeastern corner of Tennessee (Carter and Unicoi). In North Carolina, where prime habitat abounds, this species is blatantly absent except in the northwest corner, mostly in counties bordering the Tennessee sites. This species has not been recorded from Kentucky.

Look for pad-leaf orchids in areas with cooler climates and in middle to upper elevations of acidic woods, often associated with eastern hemlock or rhododendron. They particularly thrive in dark, mature hemlock woods. In forest habitats of the North, pad-leaf orchid is often found in the beech-maple hardwood zone. Though the plant's chosen habitat is dark woods, occasionally one can see its whitish raceme of flowers along the forest edge when the sunlight is just right. I once amazed one of my friends by spotting a plant from my car while going down the highway at fifty miles per hour.

Platanthera peramoena
Asa Gray

Purple fringeless orchid

The purple fringeless orchid is highly regarded as an especially beautiful wildflower. The *Flora of West Virginia* stated that the specific name for this species, *peramoena*, means "very beautiful." Luer's *Native Orchids of the United States and Canada* declared that the word comes from the Latin "peramans" and means "very loving." For me, both meanings apply. It is very beautiful, and I love looking at it.

In the prettiest color form, all parts of the flowers of purple fringeless orchid are "bubble gum" pink. Yes, more pink than purple, at least not purple as in the color of the purple fringed orchids. To some folks it is a shock-

A close-up look reveals the elegance of the purple fringeless orchid flower.

ing pink. They are shocked at the bright, alive beauty. The lateral sepals are reflexed, thrusting out the two pollinia from beneath the flattened hood or bonnet formed by the dorsal sepal and lateral petals. The lateral petals are narrow at the base and rounded on the end. The three-lobed lip is spread openly, with a small notch in the middle at the tip of the center lobe. A finely serrated (not fringed) edging marks the terminal margins of each of the three lobes.

Healthy plants of this orchid usually are about 20 inches (approximately 50 centimeters) high and typically have several dozen flowers. But I have seen plants 3 feet (approximately 90 centimeters) tall, bearing well over a hundred flowers crammed tightly into the inflorescence. When all the flowers are open, the raceme is a cylindrical shape and covers the top third of the stem. It is common for all traces of the plant to disappear quickly after flower-

Opposite:
Purple fringeless orchid is often found along fencerows in wet fields and meadows.

Left:
Typical flower racemes of the purple fringeless orchid are densely clustered with dozens of flowers.

ing. Purple fringeless orchid often grows in company with tall phlox, *Phlox paniculata* L., which greatly resembles the orchid in color, in the shape of the inflorescence, and in its choice of habitat.

Purple fringeless orchid blooms from mid-July into the first part of August. The species ranges from the central part of the Midwest to central Pennsylvania and then reaches southward across the Appalachians to northern Georgia. This is a relatively limited range. In the southern mountains, sites are fairly scarce and widely scattered. Sparse locations in the mountains of southwestern North Carolina and a few sites in Kentucky illustrate the scarcity of known populations. Ten West Virginia mountain counties have sites. There are several sites on the Cumberland Plateau but only two eastern Tennessee locations. Virginia has the largest number of sites for this handsome orchid, with populations in the James River drainage of the central highlands and in at least ten counties from the New River Valley westward. Oddly, as do several other orchid species, this orchid virtually skips over the Roanoke River drainage in the central part of the mountain highlands of Virginia.

When compared to other species in the southern mountains within the genus *Platanthera*, purple fringeless orchid appears to be in the middle as far as size of populations is concerned. It never seems to form really large colonies as does the yellow fringed orchid, *P. ciliaris*, but certainly it can be more prolific than the ragged fringed orchid, *P. lacera*.

This orchid prefers wet fields and meadows and roadside ditches. Many meadows where this plant once flourished have been developed or cleared for livestock, and the plants have been forced to exist as meager populations on the edges. Thus pushed to the edges, purple fringeless orchid has become especially vulnerable in its roadside setting to mowing, destruction by road crews, chemicals sprayed in the summer, and salt thrown by snow scrapers in the winter. Purple fringeless orchid is rapidly losing its habitat.

I often refer to this orchid as the "southern belle." The flower's pollinia resemble big, dark, enchanting eyes. The lip is spread like outstretched arms beckoning above a wide, full skirt. All this is topped off by a delicate pink bonnet that never fails to remind me of a beautiful young debutante all decorated for her first cotillion. I have concluded that the name purple fringeless orchid is totally inappropriate and lacking in imagination. Something like southern belle orchid seems to fit much more suitably for this delicate beauty.

I also call purple fringeless orchid the "hard luck" orchid. Almost every good site I have found for this plant, for one reason or another, has suffered some calamity that has either obliterated the site or caused a great reduction in population. One site has been destroyed by the introduction of cattle into the wet meadow. Another site was ravaged by, of all things, the construction of a fence to accommodate a herd of American bison. Still another site gets sprayed annually by insecticides, and it is only a matter of time until its population totally disappears. One other site in North Carolina seems to be attacked each summer by a group of beetles, which often eat through the buds before the plant can flower. But the strangest habitat alteration story I know involves a site that was "rearranged" by a young man who secretly cultivated a crop of marijuana plants among the alders and tall weeds of the purple fringeless orchid habitat.

The purple fringeless orchid is simply one of our prettiest wildflowers. It is easily the most enchanting orchid of the late summer in the southern Appalachians.

Platanthera psycodes (L.) Lindley

Small purple fringed orchid or butterfly orchid

The specific name for this lovely orchid, *psycodes*, is both interesting and quite appropriate. In mythology, Psyche was a beautiful maiden depicted as a winged fairy. Ancients believed the soul was represented by winged creatures such as insects or birds. Psyche consequently evolved to personify the soul or one's inner self. Thus we have the modern English word, "psyche," referring to a part of our existence that is in harmony with but independent of the body. In entomology, or the study of insects, there is a family of medium-sized moths called Psychidae. "Psyche" in Greek also means "butterfly" or "moth" and gives rise to butterfly orchid as one of the common names for this species.

Flowers of the small purple fringed orchid are smaller than those of its close relative, the large purple fringed orchid, *P. grandiflora*. (Compare the flower description of this species with that of the large purple fringed orchid above.) The small purple fringed orchid is usually varying shades of lilac purple but can be pinkish or even, very rarely, white. The lateral sepals of small purple fringed orchid flowers are acutely reflexed (as might be the wings of an insect) while the lateral petals are virtually upright, which tends to thrust the pollinia outward in invitation to pollinators. The lip is three-lobed, with fringe along the end margins of each lobe. The shape of the opening to the nectary is diagnostic in this species. The opening is influenced by a tiny swelling of tissue in the center that makes the opening dumbbell-shaped. The nectary opening in the large purple fringed orchid is large and round.

In prime habitat, the height of the small purple fringed orchid plant can reach nearly 2 feet (approximately 60 centimeters), often as tall as the normal height of large purple fringed orchid. Flower racemes are typically cylindrical and may have from about a dozen to many dozen tightly clustered flowers.

The cylindrical raceme is typical of the small purple fringed orchid.

In the southern Appalachians, the small purple fringed orchid normally begins to bloom in mid-June, a few days earlier than the large purple fringed orchid. Many plants are in bloom on the roadside along the Blue Ridge Parkway north of Asheville, North Carolina, by this time. This orchid seems to have a short season in the southern mountains. By early July few fresh plants are to be found anywhere other than in the very highest elevations.

This species is found in the upper Midwest, across the Great Lakes area, through Ontario, and into the Maritimes. The range then is extended southward through the Appalachians to northern Georgia. Over much of its range but more commonly in the northern part, small purple fringed orchid often hybridizes with the ragged fringed orchid, *P. lacera*, producing the tremendously variable Andrews's fringed orchid, *P.* ×*andrewsii*.

In the southern Appalachian Mountains, the small

The dumbbell-shaped opening to the nectary is a distinguishing character of small purple fringed orchid flowers. Note the pollen from another flower already deposited on the stigma.

purple fringed orchid is well represented within several counties in the southwestern mountains of North Carolina and within a half dozen counties of eastern Tennessee along the North Carolina border. It is widespread in the mountains of Virginia, but in West Virginia, even though huge amounts of suitable habitat are present, sites are known from only two counties, Pocahontas and Hampshire. The plant is not known from Kentucky.

Small purple fringed orchid is uncommon in the southern mountains, but nevertheless large populations are sometimes seen. This species prefers areas with cooler climates and higher elevations, often Canadian zones. Several hundred plants growing near the top of Mount Mitchell are up as high as one can get in the eastern United States.

Seeps in the edges of woods or wet banks on roadsides near northern hardwoods or spruce forests are the preferred habitat of small purple fringed orchid. A sunny, wet glade tucked away within a grove of red spruce or Fraser fir trees is an ideal place to find this species in the southern Appalachian Mountains.

Pogonia Jussieu

Pogonia ophioglossoides (L.) Jussieu

Rose pogonia or snake mouth orchid

Only one species represents this genus in eastern North America. *Pogonia* was once included under the genus *Arethusa* and was listed by Linnaeus as *Arethusa ophioglossoides*. The current name of this genus is taken from the Greek word "pogonias" meaning "bearded." The specific name is a combination of the Greek words "ophis," which means "snake," and "glossa," which means "tongue." Thus one common name for this species is snake mouth orchid. Both scientific names are appropriate. The lip projects forward as does the tongue of a snake. The margins and particularly the end of the lip are bearded with fleshy, watermelon red bristles. The more often used common name, rose pogonia, is obviously in reference to the lovely rose color of the flowers.

A solitary flower sits atop the stem of the rose pogonia. Although on rare occasions plants have two flowers, I have never seen two flowers open on the same plant at the same time. The lateral petals and sepals are the same rose color, the same size, and the same general shape. The three sepals are spread widely in three directions. The two lateral petals, however, are projected forward, forming a cover for the pollen-bearing column. The lip of rose pogonia is a beautifully fashioned work of art. The bristles, or beard, cover the edges but also extend onto the top surface of the lip. The rose-colored bristles become shorter toward the base of the lip and gradually change color from rose to yellow and green. An insect following the trail of pollen-promising bristles ends up at the base of the lip, where it encounters a proportionately large column. To escape, it then has to brush by the pollinia, picking up pollen to distribute to other flowers.

Plants of rose pogonia are typically about 8 inches (approximately 20 centimeters) high, but some, in ideal situations, may be a foot (approximately 30 centimeters) tall. There is a solitary somewhat oval-shaped leaf sheathing the stem just above ground level.

This orchid species blooms in the southern Appalachians about the last week of June. The continental range

A rare plant of rose pogonia with two flowers on the same stem grows in a West Virginia mountain bog.

of rose pogonia extends from Newfoundland southward all along the coastal plain to southern Florida and westward across Ontario to the western Great Lakes and across the upper Midwest. In the southern mountains, this is a rather uncommon species. In North Carolina, sites are known in less than a half dozen counties. Only two counties in the eastern Tennessee mountains have sites, although several counties on the Cumberland Plateau contain sites for this orchid. Three counties in the eastern Allegheny Mountains of West Virginia have the rose pogonia, the best-known of which is Pocahontas County, the home of the Cranberry Glades. There are records from three mountain counties in Virginia, the most recent discovery being in Washington County in the southwestern part of the state. This orchid is not known from Kentucky.

Rose pogonia usually seeks out the acidic situations

Bright colors on the lip attract pollinators to the rose pogonia flower.

found in bogs. It almost always is found growing with grass-pink orchid, *Calopogon tuberosus*. But the reverse is not true. I know many sites for grass-pink orchid that do not have rose pogonia but nearly every site I know for rose pogonia also has grass-pink orchid. This is true from the coastal plain of North Carolina to Cape Breton Island in Nova Scotia. Both species are plentiful along the boardwalk at the Cranberry Glades around the first of July. Just a few miles away, however, at Droop Mountain Bog, there are plenty of healthy grass-pink orchid plants, as well as several other orchid species, but not one rose pogonia is to be found.

Shadow witch orchid plants are green and sparsely flowered, which makes them difficult to locate in their habitat.

This genus was named for French botanist Henri de Ponthieu. In the late eighteenth century, Ponthieu wandered the area of the Caribbean collecting plants, and most likely he came upon the shadow witch orchid during his travels. This species, once listed under *Arethusa*, now is the only member listed in North America for the genus *Ponthieva*. The specific name, *racemosa*, means "raceme," which is the type of inflorescence seen in the shadow witch orchid.

Flowers of the shadow witch orchid are green. Unlike most orchid species in the United States, the shadow witch orchid has the lip at the top of the flower. The lip is greatly reduced in size and rolled into a rounded bun shape. The pollinia are bright yellow and narrow, looking something like tiny bananas. The wide lateral petals are spread at the bottom of the flower and are an attractive light green with darker green venation.

Ponthieva
R. Brown

Ponthieva racemosa (Walter) Mohr

Shadow witch orchid

The lip is uppermost (nonresupinate) and reduced in size in the flower of the shadow witch orchid.

The entire plant, except for the basal rosette of dark green leaves, is sparsely covered with short, coarse hairs. The reddish-purple stem is typically about 10–12 inches (approximately 25–30 centimeters) high, with the flowers of the raceme standing well away from the stem. The inflorescence of shadow witch orchid is similar in appearance to, but has fewer flowers than, the fringed orchids, members of the genus *Platanthera*.

Shadow witch orchid can barely be considered present in the southern Appalachians and, were it not for one site on the very edge of the area covered by this book, would not be included here. The lone site, in central Tennessee, is in an open limestone seepage area with a few scraggly red cedar trees, some redbud and yellow poplar, and several sycamore trees more or less ringing the seep. The seep is open to the sun, but the orchids stay pretty close to the edges, among the trees, where they are better shaded.

The blooming time for this species, whether at the edge of the mountains or on the coastal plain, is about September 10. I have experience with coastal sites in the area of Williamsburg, Virginia, and near New Bern, North Carolina, and each one comes into bloom at the very same time as does the site I visited in Warren County on the Cumberland Plateau in Tennessee.

Coastal plain sites for shadow witch orchid are often in dark, moist areas, where the orchids grow among cypress "knees." Cypress knees are rounded roots that protrude from the ground (or water) and look like bent knees, such as those seen in bald cypress trees, *Taxodium distichum* (L.) L. C. Richard.

Spiranthes L. C. Richard

The ladies' tresses

The name *Spiranthes* means "coiled" or "spiraled." The group or genus of orchids named ladies' tresses is so called because of the spiraling pattern of the flowers as they ascend the stem. This pattern is reminiscent of the braided hair or "French braid" that many women and girls employ when styling their hair. Although all species of this genus in North America have flowers that spiral around the stem in one or more ranks, some species also have flowers arranged more or less in a line on only one side of the stem. Such an arrangement is described as secund.

There are nearly two dozen species within this genus across the United States and Canada, excluding species that are restricted to Florida. In the area of the southern Appalachians covered by this book, eight species of ladies' tresses are listed. This is the second largest genus in the southern mountains, second to the fringed orchids, *Platanthera*.

Leaves of the ladies' tresses are of two types: those that are present when the plant flowers and those that wither prior to blooming. The latter leaves are called "fugacious" leaves.

All of the ladies' tresses in the southern mountains have very small, white or off-white flowers that are superficially similar. Some species have well-defined coloring on the lower petal, or lip, that distinguishes that species. But in some species of ladies' tresses, differentiation is much more subtle.

For identification purposes, one can look at the amount

and type of pubescence, or hairiness, on the plant. This can be accomplished in most cases with the use of a good hand lens or, in the lab, a stereoscope. Ladies' tresses in the southern mountains have hairs of two types. Most species have hairs that are said to be capitate, which means that the hairs have a small glandular knob at the tip. Hairs of the other type are pointed on the tip and are found in only one species. Also one species of ladies' tresses in the southern Appalachians is glabrous, meaning it is without pubescence.

Spiranthes cernua (L.) L. C. Richard

Nodding ladies' tresses

The specific name, *cernua*, means "nodding." The base of the flower is arched prominently downward, giving the flower a nodding posture. This species is also called autumn ladies' tresses in deference to its blooming time.

Nodding ladies' tresses have rounded and very white flowers, although the lip has a yellowish flush on the interior because of its thickness. The outer part of the lip has a translucent quality not seen in the other flower parts.

Plants of nodding ladies' tresses may be up to a foot (approximately 30 centimeters) high, with several very narrow basal leaves present at the time of flowering. The flowers are arranged in more than one rank, and each rank spirals upward around the top half of the stem. The lower part of the plant may be smooth, but the entire upper part is sparsely covered with fine capitate hairs.

The species known as nodding ladies' tresses is the base for a complex of ladies' tresses species that once included yellow ladies' tresses, *S. ochroleuca*; Great Plains ladies' tresses, *S. magnicamporum*; and the coastal fragrant ladies' tresses, *S. odorata* (Nuttall) Lindley; all of which were once listed only as varieties of nodding ladies' tresses. Each has now been described as a separate species by one authority or another. Nodding ladies' tresses are often found growing alongside their close cousins, the yellow ladies' tresses. In these situations, nodding ladies' tresses will be in lower, wetter areas, while the yellow ladies' tresses are on higher ground and in a drier location, often in the margins of woods. Nodding ladies' tresses is the most common species of the genus *Spiranthes* in the eastern United States. It often forms large colonies that can number hundreds of plants in wet

*Above:
The nodding posture at the base of the flowers is typical in nodding ladies' tresses plants.*

*Left:
A large population of nodding ladies' tresses grows in a wet roadside meadow.*

meadows and roadside ditches. In total population numbers, nodding ladies' tresses is second only to downy rattlesnake plantain, Goodyera pubescens, as the largest orchid species in the southern Appalachian Mountains.

This orchid has a wide blooming range. It can be found in flower in lower elevations in late August, but, in higher elevations like the Cranberry Glades of West Virginia, fresh plants can be found in November. In most areas of the southern mountains, nodding ladies' tresses reaches its peak of flowering in mid-September, when the birds of prey are making their incredible migratory flights across the Appalachian ridges on their way to South America for the winter. This is a favorite time of year for me. I can sit on a mountaintop along the Blue Ridge Parkway watching the spectacular flights of hawks, eagles, and falcons and enjoy the nodding ladies' tresses in the roadside ditches at the same time.

Opposite:
The southern variety of slender ladies' tresses has a tight spiral of flowers ascending the stem.

Left:
The northern variety of slender ladies' tresses has leaves present at the time of flowering.

As with the ragged fringed orchid, *Platanthera lacera*, the specific name for this species refers to cutting or laceration. In this case, "lacera" refers to the minuscule lacerations on the edge of the lip.

The flowers of slender ladies' tresses are tiny, about one-eighth to one-quarter inch (approximately 8 millimeters) long. A green lip, marginally edged with white, distinguishes this species. The pearl white lateral petals and dorsal sepal, along with the lip, form the tubelike flower, whose lateral sepals are somewhat spread. Slender ladies' tresses plants may be anywhere from about 8 inches to 15 inches (approximately 20–38 centimeters) tall.

This species has two varieties that are now recognized as distinct by most orchid authorities. The identical flowers of both varieties can cause a great deal of confusion, but knowing the differences in the leaves, the geographic location, the time of flowering, and the overall appearance of the plant can eliminate most of the problems in separating the two varieties.

Spiranthes lacera (Raf.) Rafinesque

Slender ladies' tresses

Below, left:
Spiranthes lacera *(Raf.) Rafinesque var.* gracilis *(Bigelow) Luer, southern slender ladies' tresses*

Below, right:
Spiranthes lacera *(Raf.) Rafinesque var.* lacera, *northern slender ladies' tresses*

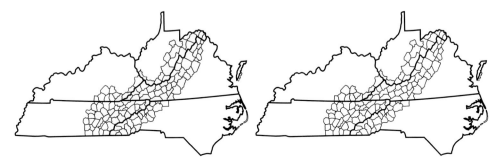

Spiranthes lacera
var. *gracilis*
(Bigelow) Luer

Southern slender ladies' tresses

The more southern variety is named var. *gracilis* (Bigelow) Luer, *gracilis* simply meaning "slender." This variety extends its range westward to the Great Plains and northeastward to the Great Lakes and New England. It overlaps somewhat in range with its northern counterpart, but very little of the overlap occurs in the southern Appalachians. The southern variety is found in the same type of habitats as the northern variety: old fields, roadsides, cemeteries, and forest margins. It even pops up in the grass of yards that haven't been mowed for a week or so. This southern variety is one of our more common species of orchids in the southern mountains.

The smaller flowers of the southern variety of slender ladies' tresses are very tightly crowded into a spiral that may contain two or three ranks but usually contains just one. The flowers are inconspicuous, but an exacting investigation of them will unveil the finest of woven interlacing, an intricacy that reminds me of my wife's stitching on her finest crochet projects. The leaves wither before flowering, and the plant is only slightly pubescent or may even be considered smooth. This variety of slender ladies' tresses is in bloom from late August into September.

Spiranthes lacera
var. *lacera*

Northern slender ladies' tresses

The two varieties of slender ladies' tresses are at least partially separated by range. The typically more northern and western variety, which has been given the name var. *lacera*, does not come into the range of the southern Appalachians as defined by this book except in the very localized area where Kentucky, Tennessee, West Virginia, and Virginia meet in the Cumberland Mountains. And even there, few locations are known. This variety inhabits fallow fields, forest glades, and roadsides in rather dry situations.

The varietal differences in regard to plant shape and flowering are a bit more profound than the differences in range. The northern variety is a somewhat larger plant, and its naturally larger flowers are more widely spaced along the stem than those of the southern variety. The plant is also rather pubescent, with fine capitate hairs. The flowers are spiraled about the stem in a single rank. Leaves are formed in a basal rosette and are usually still prominent when the plant blooms. This northern variety blooms in mid-July, some three to four weeks before the appearance of the southern variety. Fred Case, in *Orchids of the Western Great Lakes Region* (1987), noted that the northern variety of slender ladies' tresses "abounds in open Jack-pine and Scrub Oak forests," indicating that, while it may be very rare in the southern mountains, in its more northerly situations it can be quite abundant.

The species name for this orchid is taken directly from the Latin word "lucidus," meaning "clear," as in a lucid explanation. But "lucidus" can also mean "bright" or "shining," which in this case refers to the polished or shining appearance of the leaves of this orchid, which shine like waxed cucumbers in produce markets.

Flowers of shining ladies' tresses are quite small and tubular. The bright white sepals and petals do not spread but flare slightly at the tip. In contrast to this white coloring, the lip is a striking lemon yellow. This yellow coloring easily distinguishes the species from any other in the genus.

The glossy leaves are present at flowering and are somewhat shorter and wider than those of most other species in this genus. Plants found in the southern mountains are usually no more than 6–8 inches (approximately 15–20 centimeters) tall but in some situations in the North may reach nearly a foot (approximately 30 centimeters). The spike of flowers is a spiral of three-ranked blossoms. The flowers are congregated more toward the top of the stem than is the case in most species of ladies' tresses, which makes the spike comparatively shorter. The plant's appearance is usually slender and dainty, but the bright yellow coloring on robust plants can be quite attractive. The lower parts of the plant are smooth, but the flowers and bracts are slightly pubescent with capitate hairs.

Shining ladies' tresses flower in late May or early June in the southern Appalachian Mountains, about a month and a half before any of the other members of this genus. This species is known to grow primarily in more basic soils but does rarely grow in rather acidic situations.

The overall range for shining ladies' tresses is across the upper Midwest and lower eastern Canada through New

Spiranthes lucida (H. H. Eaton) Ames

Shining ladies' tresses

Shining ladies' tresses is a rare species in the southern Appalachian Mountains.

England into the lower Maritimes. This orchid is at the southern limit of its range, and is rarely encountered, in the southern mountains. The few and sparsely scattered sites in West Virginia have been enhanced by a 1999 discovery in Monroe County, which brings the state's total to five counties with records for this species in the eastern Alleghenies covered by this book. In Virginia there are several old records for the northern Shenandoah Valley and one lost site in Smyth County in the southwest. Recent discoveries of this species in Giles, Pulaski, and Montgomery Counties offer the hope for more Virginia locations in the future. Shining ladies' tresses have never been seen in North Carolina. Luer did not include Tennessee in his range map for shining ladies' tresses in *The Native Orchids of the United States and Canada*. Nevertheless, the state does have several widely scattered sites. One new site discovered in 1996 in Overton County, west

The bright, lemon yellow lip distinguishes shining ladies' tresses from all other species in its genus.

of the Cumberland Plateau, is said to have a large population. Despite the fact that Luer did include the eastern half of Kentucky in his range map, shining ladies' tresses has never been listed as a species actually found in the state.

Sometimes a population of shining ladies' tresses will lie dormant for several years before a single plant emerges to bloom. A site that is still barely viable near my home in southwestern Virginia went from 1990 to 1995 without displaying a single plant. A Claiborne County, Tennessee, site has very few plants, but they are particularly robust and quite beautiful. This site on a steep Tennessee hillside also has showy lady's slipper, *Cypripedium reginae*, growing in a limestone seep. With the presence of both of these species of particularly rare orchids, this has become a highly prized location.

Lateral sepals often spread widely on the flowers of Great Plains ladies' tresses.

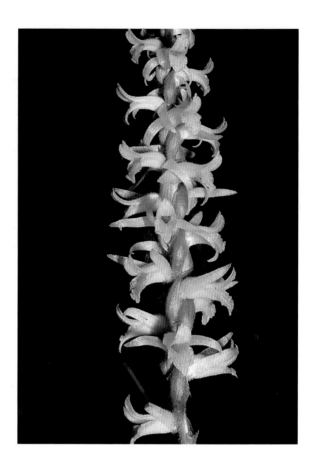

Spiranthes magnicamporum Sheviak

Great Plains ladies' tresses

The specific name, *magnicamporum*, comes directly from the Latin "magnus," meaning "large" or "great," the same word from which we get "magnificent" or "magnitude," and "campus," meaning "field" or "plain," as in the campus of a university. In the case of this orchid species, the two Latin words refer to the Great Plains of the United States, where it is common.

The flowers of Great Plains ladies' tresses are white or off-white. The lip is oval-shaped, with none of the dilation at either end that is seen in the similar nodding ladies' tresses, *S. cernua*. The center of the lip is a pale yellow. Lateral sepals usually separate from the other flower parts and curve inward and upward. The flowers have a sweet fragrance.

Great Plains ladies' tresses plants have fugacious leaves and a stout stem with a crowded spiral of flowers arranged in three or four ranks. Particularly robust plants may be 15 or more inches (approximately 38 centimeters) tall.

In the southern Appalachian Mountains, plants in the only known area for Great Plains ladies' tresses begin flowering in late September but really get into full swing in October. This area consists of two or three sites, all in Russell County, Virginia. Needless to say, occurring in only one known area, this is an extremely rare species in the southern mountains. Other fairly recent discoveries of this orchid have been made near the southwestern Cumberland Plateau and in Georgia just south of the Tennessee border.

Typical habitat for the Great Plains ladies' tresses is in areas with definitively calcareous soils. Cedar glades seem to be its habitat of choice in areas of north Georgia near the southern mountains. It is no surprise, then, that the one area in the southern Appalachian Mountains for this orchid would also be made up of sites that are particularly calcareous. Specifically, a site that I have visited is on a very steep slope of barren, rather dry limestone, much like an ancient prairie that has been lifted into the mountains.

The finding of Great Plains ladies' tresses in southwest Virginia in 1996 was a particularly significant event. Not only unknown from the mountains, the species was unknown from the entire area made up of the states of Virginia, West Virginia, Kentucky, North Carolina, and Tennessee (although there are rumored to be a couple of sites in central Tennessee that are thus far undocumented). The Virginia sites place the species about midway between its known traditional range in lower Ontario and the lower Great Lake states to the north and the upper areas of the Gulf states to the south. This would suggest that there are likely more areas within this vast undocumented region where Great Plains ladies' tresses might be found.

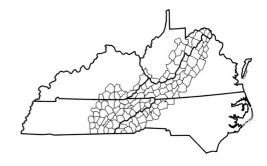

Yellow ladies' tresses plants are often found in bloom as late as November in the southern Appalachians.

Spiranthes ochroleuca (Rydberg) Rydberg

Yellow ladies' tresses

Because of the yellowish appearance of the flowers of this species, it was given the name *ochroleuca*. The name was taken from the Greek "ochros," meaning "pale," and "leucos," meaning "white." When compared to the flowers of its close relative, nodding ladies' tresses *S. cernua*, the flowers of this species are usually more slender and have a more yellowish color, and they are oriented either horizontally or upward on the stem rather than nodding.

The plants of this species are smooth on the lower parts but have capitate hairs on the bracts and flowers. Leaves of yellow ladies' tresses are long and narrow, emanating from the base of the stem, and are present when the flowers open. Healthy plants of yellow ladies' tresses may be 12–15 inches (approximately 30–38 centimeters) high and are found in drier habitats of banks, fields, and forest edges. This species blooms from September to November, a bit later than nodding ladies' tresses, but the bloom times of the two species do overlap in much of their range.

This orchid has previously been considered a more northerly ranging species. In 1975, Luer described the range of yellow ladies' tresses as reaching from the upper Great Lakes across to lower New England, Nova Scotia, and Newfoundland. His range map showed no locations south of lower Pennsylvania.

But since the publication of Luer's book, many new sites have been located, especially in the high mountain country of West Virginia and in the mountains of southwest Virginia. In 1970, when the revised four-volume *Flora of West Virginia* was published, yellow ladies' tresses was not recognized as ever having been found in the Mountain State. By 1996, however, this species had been documented in no less than 29 counties in West Virginia, 12 of which are within the range of this book. The original *Atlas of the Virginia Flora* listed no counties for this species in 1977. But by the publication of the third edition of the same atlas in 1992, five mountain counties in northern Virginia were listed. Since 1993, I have myself found yellow ladies' tresses at new sites in Craig, Giles, Bland, Pulaski, Wythe, Smyth, and Grayson Counties in the southwest mountains of Virginia. This block of counties extends the range of the orchid solidly from the West Virginia border across southwest Virginia to the North Carolina border. One location is as far south as the area around Boone, North Carolina. Two locations are known in Tennessee, one on the Cumberland Plateau and one in the eastern mountains in Sevier County. There is no mention of this plant in Ettman and McAdoo's *Kentucky Orchidaceae* (1978).

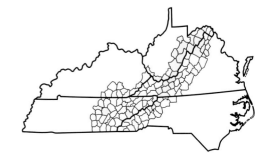

The name oval ladies' tresses refers to the shape of the raceme of flowers.

Spiranthes ovalis Lindley var. *erostellata* Catling

Oval ladies' tresses

The specific name, *ovalis*, comes from the Latin "ovum," meaning "egg." This word is the basis for several modern English words: "ovary," the producer of eggs; "ovipositor," the egg-laying apparatus of insects; and even "oviraptor," a type of dinosaur that preyed on others by eating their eggs. And of course it is the origin of "oval," meaning "egg-shaped," as reflected in the species name of this orchid, which makes reference to the shape of the inflorescence, which is smaller on either end.

Plants of oval ladies' tresses in the southern Appalachians belong to var. *erostellata*. Plants of this variety are small and lack a viscidium; thus they cannot attach their pollinia to a pollinator and rely on self-pollination.

This orchid species blooms in September in the southern mountains. The stem of oval ladies' tresses is usually 8–10 inches (approximately 20–25 centimeters) high. The short inflorescence may have several dozen tiny white flowers packed tightly in several ranks at the tip of the stem. The end of the lip is somewhat more narrow or pointed than in the other ladies' tresses species. The plant is pubescent, with capitate hairs, and two to three narrow leaves are present at flowering. Oval ladies' tresses are true forest dwellers and do well in the shade. Most situations where the plants are present in the southern mountains are in rather dry calcareous soils, but they may also thrive in other situations.

Although Luer's range map for oval ladies' tresses in *Native Orchids of the United States and Canada* (1975), showed this species as distinctly southeastern, several well-established locations were actually found in the late 1960s and 1970s in Michigan, Illinois, Indiana, Ohio, and even Ontario. In *Orchids of the Western Great Lakes Region* (1987), Fred Case suggested, "It may be expanding its range, or be more widely present than had been suspected."

Michael A. Homoya, in *Orchids of Indiana* (1993), described some plants of oval ladies' tresses with flowers that do not open yet reproduce themselves very well. This self-pollination property of cleistogamous flowers (discussed under Bentley's coralroot orchid, *Corallorhiza bentleyi*, above) helps to explain how the range of oval ladies' tresses might have expanded into the North, where natural pollinators may not be present.

Oval ladies' tresses locations are quite limited in the southern Appalachian Mountains. There are six counties with records from Virginia's mountains. But no one seems to know of a recently viable location. West Virginia has one extant site in the mountains in Jefferson County very near the Potomac River, a lower-elevation area. In Tennessee, only two counties (Sullivan and Sevier) in the eastern mountains have records for this species, but there are nearly a half dozen sites recorded from the Cumberland Plateau. Kentucky lists four counties for oval ladies' tresses in the Cumberland Mountains (Bell, Harlan, Letcher, and McCreary). North Carolina has no mountain sites for this orchid. The small size of the plants often causes them to be concealed, which probably contributes to the scarcity of known sites.

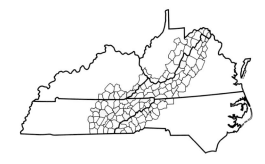

Little ladies' tresses flowers have small, tubular, all-white flowers.

**Spiranthes
tuberosa
Rafinesque**

**Little ladies'
tresses**

The common name for this species obviously comes from the small size of the flowers. The scientific name, *tuberosa*, is derived from the Latin word meaning "swollen" and refers to the bulging singular tuber. Some authorities still use the scientific name *S. grayi* Ames, which, of course, is in reference to the famed botanist Asa Gray.

Flowers of little ladies' tresses are tube-shaped, pearly white, and very small. The diminutive size is a factor in this species's being considered more rare perhaps than it really is, since it means that the plants are often overlooked. The lateral sepals spread somewhat but not as much as in other ladies' tresses species. Petals do not flare, which helps give the flowers their tubular shape. The lip is very white, with only a hint of pale yellow in the center. The plant is very slender, although individuals may reach a foot (approximately 30 centimeters) in height. This species is completely glabrous, lacking the pubescence of the other species in the genus. Several small, oval-shaped leaves will persist over winter but wither prior to the time of bloom. Little ladies' tresses begin to flower in the southern mountains early in August in some areas, even though some sites remain prime into September.

This orchid species is spread widely across the southeastern United States except for southernmost Florida. Most locations are situated, however, on the coastal plain or in the piedmont, well away from the mountains. Little ladies' tresses can be found in the southern Appalachians primarily in the areas of the Cumberland Mountains in Tennessee and Kentucky. Nearly half the counties on the Cumberland Plateau of Tennessee have records for this plant. I have seen little ladies' tresses near the parking area at the spectacular Fall Creek Falls in Van Buren County. A few eastern counties of Tennessee also have scattered sites. North Carolina has a few mountain locations for this species in the extreme southwestern corner of the state, and three mountain counties in Virginia report the presence of little ladies' tresses. There is a lone record for West Virginia, from Hampshire County in the northeastern panhandle of the state. These records indicate a southern distribution mainly in areas with close affinities to the lower-elevation sites where the orchid grows outside the mountains. Although not considered particularly rare, this orchid is found uncommonly in the southern mountains. It does have a fairly wide range, but the number of sites within the mountains is rather limited.

Little ladies' tresses is known primarily from well-drained or dry areas, such as the edges of pine-oak forests, open fallow fields, and bare, rocky roadsides. This species seems to care little about "keeping its feet wet," as is the habit of most members of this genus.

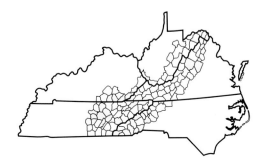

Spring ladies' tresses plants vary widely in size and, oddly enough, bloom in late summer in the southern mountains.

Spiranthes vernalis Engelmann and Gray

Spring ladies' tresses

We are most familiar with the word "vernal" from awaiting each year the vernal, or spring, equinox, which officially marks the beginning of spring. The word is taken directly from the Latin "vernalis," which means "spring." As the species name for this orchid, *vernalis* refers to its blooming time in much of its coastal range. In the southern Appalachians, however, spring ladies' tresses do not bloom until late July and early August, and in some colder climate areas, not until September.

Flowers of spring ladies' tresses are pale white and widely variable in size. The lip has a yellowish center and is usually open and downward curving, in opposition to the dorsal sepal and lateral petals, which are open and somewhat upward curving. One good field character for identification of this orchid is the pubescence, which has sharply pointed hairs, the only species in the genus in the southern mountains with pointed hairs.

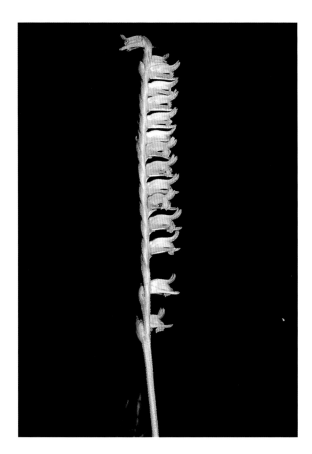

This plant of spring ladies' tresses exhibits secund flowers arranged in a row on one side of the stem.

The size of spring ladies' tresses plants can vary a great deal also. Some plants are slender, with smaller flowers, and no more than 8–10 inches (approximately 20–25 centimeters) tall. But some plants exceed 3 feet in height and have rather large flowers. I once found a specimen in Wythe County, Virginia, that was 40 inches (approximately 1 meter) tall and had two neighbors that were each over 2 feet (approximately 60 centimeters). Flowers are usually spiraled around the stem but may also be secund, or arranged on only one side of the stem. Four or five linear, rather stiff leaves are usually present at blooming time.

This species ranges from the eastern Great Plains to the Gulf of Mexico and eastward to the Atlantic shore. It skips the upper Midwest but does hug the coast as far as lower New England. In the southern mountains, locations are widely scattered. Spring ladies' tresses was not listed for

West Virginia in Strasbaugh and Core's flora of 1970 but now has been found in four counties, although only one of these, Barbour County, is in the Allegheny ridges of eastern West Virginia that are within the scope of this book. There is only one county record in southeast Kentucky, in Bell County. In Virginia, no mountain counties were listed for spring ladies' tresses in the 1977 issue of the *Atlas of the Virginia Flora*. By the publication of the 1992 edition, the species had been verified in seven mountain counties of the Old Dominion, and I have since found additional sites for this species in Giles, Grayson, and Wythe Counties. Locations in North Carolina's mountains are particularly sparse, but I suspect there are many more sites that are still unknown. Tennessee seems to have the majority of the mountain locations for spring ladies' tresses in the southern Appalachians, with records from 12 different counties on the Cumberland Plateau and along the eastern mountains that border North Carolina.

The habitat for spring ladies' tresses in the southern mountains is usually along roadsides and in dry fallow fields. I have found this species in several areas of the Blue Ridge Parkway on both sides of the border between North Carolina and Virginia. But the species is also reported from the boggy environs around Long Hope Bog in the high mountains of North Carolina.

Tipularia Nuttall

Tipularia discolor (Pursh) Nuttall

Crane-fly orchid

The flowers of this orchid resemble a crane fly, giving rise, obviously, to the common name and also to the genus name, *Tipularia*, taken from Latin "tipula," which connotes the insect known as the crane fly. The specific name, *discolor*, comes from the prefix "di" meaning "two," and, of course "color." The two colors referred to are those of the leaf, which appears in the winter and is dark green on the top and bright purple or beet red on the bottom.

The sepals and petals of crane-fly orchid flowers are long, narrow, and widely spread, looking every bit like the insect. Whether brownish, greenish, or a subtle purple color, all parts of the flowers on any given plant are the same color, except for the column, which is a luminescent green that puts me in mind of the business end of a lightning bug.

The solitary, bicolored leaf of crane-fly orchid appears

Above:
The hibernal or winter leaf of the crane-fly orchid has beet-red coloring on the back.

Left:
Crane-fly orchid plants flower in the late summer in dark woods and often go undetected.

in early winter but withers away completely by bloom time. The leaf has heavy parallel venation and may often have a rough texture, with small "bumps" appearing on the surface. There are three species in this genus worldwide, but only this one occurs in North America. All three species have a hibernal, or winter, leaf. The flower stalk may be 15–18 inches (approximately 38–46 centimeters) high, with the top half formed into a slender cylindrical raceme of loosely spaced flowers.

This species blooms around the first of August. It appears most often in rich leaf duff in well-shaded, mature woods. Crane-fly orchid is tolerant of both acid and more basic soils and thus is found in many types of woods situations. This species often displays a partiality for woods with plenty of American beech trees, *Fagus grandifolia* Ehrhart.

The northern range limit for this orchid is mid-Ohio,

A detached pollinia from this crane-fly orchid flower has adhered to one of the lateral petals.

northern West Virginia, and from Maryland along the Atlantic Coast to southeastern New York State. Crane-fly orchid is found all across the southeastern United States except for most of Florida. In the southern Appalachians this orchid is rather common, although its presence is often unnoticed. With its slender plant, thin flowers, and subdued colors, crane-fly orchid hides easily in the well-canopied, dark woods of late summer.

Crane-fly orchid is known from virtually every mountain county of North Carolina. It is very common across the Cumberland Plateau and the eastern mountains of Tennessee. Likewise, the Cumberland Mountains of southeastern Kentucky have populations in most every county. This species appears in eight counties in the eastern Allegheny ridges of West Virginia. In Virginia, crane-fly orchid is extremely common from the Blue Ridge eastward to the coast. But the frequency of the species varies in Virginia's mountains. It is present in virtually every mountain county in southwest Virginia but glaringly absent from the records of Virginia's western Allegheny ridges.

The scientific name for the genus, *Triphora*, means "three-bearing," in reference to the three flowers that are presumed on each stem. The fact is, however, that not all stems have three flowers. Many have two or even one, and very rarely they may have four. And it is uncommon indeed to find three flowers in bloom on the same stem at the same time. Oddly enough, the specific name, *trianthophora*, also means "three-bearing." The common name is from a fanciful image of the flowers' resemblance to three birds in flight. There are a total of five species from this genus in the United States. All five are found in Florida, but the three-birds orchid is the only one found north of the Sunshine State. At various times through the years, the genus *Triphora* has been listed under the genera *Arethusa* and *Pogonia*.

The flowers of three-birds orchid are beautiful. The sepals and the lateral petals are the same color, either white or delicately lovely shades of pale purple. Occasionally some flowers display a very deep purple. The three sepals are spread but not widely. The two lateral petals arch forward over the column and are slightly curved upward at the tip. The lip is three-lobed, with the side lobes curled upward forming the rounded sides of the tubelike flower. In a freshly opened flower, the impressive lip is curved downward and has tiny green bristles along the surface, giving it a "bearded" look. The flower stems or pedicels are rather weak, and the weight causes the flowers to nod, thus another common name, nodding pogonia.

Plants of three-birds orchid are usually no taller than 8 inches (approximately 20 centimeters). The stem is usually reddish-brown but sometimes the upper part may be green. The several small, heart-shaped leaves are sessile, or stalkless, and stand separately along the stem. To me, the reduction in the size of the leaves indicates that this orchid would not require proportionately as much photosynthesis as do most other orchid species. Consequently, three-birds orchid would likely rely for much of its nour-

Triphora
Nuttall

Triphora trianthophora (Swartz) Rydberg in Britton

Three-birds orchid or nodding pogonia

ishment on another green plant by way of its relationship with its fungus partner. Three-birds orchid seems to have a habit of staying underground until enough energy is stored and the right conditions for flowering are present. This underground period may last for years and may help account for the unpredictable numbers within populations of this orchid from year to year that make the species seem so erratic in its blooming behavior.

Bloom time in the southern Appalachians for three-birds orchid is usually within the first half of August, but predicting the exact blooming time can be extremely frustrating. Flowering appears to be directly associated with fluctuations in the weather. Carlyle Luer, in *The Native Orchids of the United States and Canada,* explained: "All mature buds open at the same time on the same day. Frequently, a few degrees decrease in the night temperature is followed 48 hours later by the mass flowering of a colony, or all the colonies in that region subject to similar meteorological conditions." This temperature-related occurrence of parallel blooming is called "thermoperiodicity." Luer further noted that, "experimentally, budding plants placed for one night in a cool-house, where the temperature is about 10 degrees lower than the average outdoor night temperature, will flower two mornings later." But Luer also stated that sometimes the plants flower without the significant temperature decrease and that, "flowering is probably induced by a delicate combination of factors still unknown." New or unripened buds are unaffected by the temperature decrease, meaning that plants in a population may undergo a series of blooming cycles until all the buds have opened. Individual plants may have a wilted flower, a fresh flower, and an unopened bud all at the same time. Flowers are very short-lived, just one day. Thus there is a narrow window of opportunity for pollinators and for people who want to see them.

This orchid is found in the eastern United States as far north as New England, except for the upper Great Lakes, and as far south as the Gulf Coast, excepting the southern Atlantic Coast. In the southern mountains, locations for three-birds orchid are widely scattered and few in number. West Virginia has no site on the eastern front of the Alleghenies, but several fantastic sites are known from Barbour, Webster, and Randolph Counties. In Virginia, five mountain counties have known sites. North Carolina

Opposite:
The gorgeous flowers of three-birds orchid bloom in deep woods in late summer.

can boast of nearly a dozen counties that have population records, mostly in the southwestern section of the state. Tennessee has several counties with sites for three-birds orchid on the Cumberland Plateau and five counties with sites in the eastern mountains on the border with North Carolina. Only Bell County, in the area of the Cumberland Gap, where many good sandstone habitats are present, is listed as a location for three-birds orchid in southeastern Kentucky, but no doubt more populations exist in this section of the Bluegrass State. All things considered, three-birds orchid is probably not as rare as many people might think. Its small size, together with its rather unpredictable blooming pattern, makes it difficult to observe and photograph in its prime flowering state.

Three-birds orchid is always found in particularly acidic soil situations. Preferred habitat in the southern mountains has traditionally been described as dark hemlock woods. But more recent discoveries have shown that there are also a number of populations found in mixed hardwoods. One special West Virginia site is in a section of forest where the mature trees are oaks and maples, with a few poplars nearby. There isn't a hemlock in sight. There is, however, a large sandstone outcropping that provides the required acid. And the three-birds orchid plants at this site are primarily found among the boulders, where the soil acidity is highest.

Glossary

Anther: the part of the stamen where pollen is located.

Bract: a reduced leaf, either along the stem or at the flower base.

Capitate: shaped like a head.

Capsule: the mature ovary that splits apart to scatter seeds.

Chasmogamous: describes a normal flower that spreads its perianth to the advantage of its pollinators.

Cleistogamous: describes a flower that does not open its perianth to pollinators and typically self-pollinates.

Column: the orchid flower reproductive structure, made up of the fused stamens and pistil.

Cordate: heart-shaped.

Corm: a short, erect underground stem.

Dicotyledon: a flowering plant that has two seed leaves.

Endemic: restricted to a particular geographic area, native.

Entire: unlobed and unfringed, without division.

Filament: the threadlike upright stalk that supports the anther of the stamen in a flower.

Fugacious: describes plant leaves that wither prematurely or disappear before the plant blooms.

Genus: a group of plants or animals with like characteristics; ranked between family and species.

Glabrous: smooth, without hairs.

Hybrid: the offspring of two genetically distinct species.

Inflorescence: the arrangement of flowers on a plant.

Labellum or lip: the petal in the orchid flower that differs from the other two petals in shape, size, or color.

Lanceolate: lance-shaped; wider at or above the base but much longer than wide and tapering toward the tip.

Mentum: a small, spurlike structure, which may contain nectar, located at the base of some orchid flowers.

Monocotyledon: a plant with only one seed leaf.

Morphology: the appearance of a plant or animal with regard to form and structure.

Mycorrhiza: a symbiotic relationship between the root system of a plant and a fungus.

Nectary: the part of a flower that secretes nectar.

Ovary: the lower part of the pistil of a flower where the ovules are borne.

Ovule: an immature, unfertilized seed.

Pedicel: the flower stalk that connects the flower to the main stem.

Petal: a part of the inner whorl of floral structures that typically functions in attracting pollinators.

Pistil: the female reproductive part of a flower, made up of the ovary, style, and stigma.

Pollinarium: the plant segment made up of the pollinium, its stalk, and the viscidium.

Pollinium: a singular mass of pollen.

Pubescent: downy or hairy.

Raceme: an unbranched inflorescence of pedicelled flowers generally opening from the bottom to the top.

Recurved: arching backward.

Reflexed: arching acutely backward.

Reniform: kidney-shaped.

Resupination: the process in an orchid flower in which the opening bud turns so that the lip is the lowermost petal.

Reticulation: a network of veins or markings.

Rhizome: a horizontal underground stem.

Saprophytic: obtaining sustenance from organic matter that is no longer living.

Secund: having flowers arranged in a row on one side of the stem.

Sepal: a part of the outer whorl of floral structures that typically functions in attracting pollinators.

Sessile: having no stalk, usually referring to a leaf or a flower.

Species: the most singular ranking of a plant or animal; the division ranked below genus.

Spike: an unbranched stem with an inflorescence of sessile flowers.

Spur: a slender backward extension of the lip in an orchid flower.

Stamen: the male part of the flower, consisting of the filament and anther.

Staminode: a shieldlike or triangular non-pollen-bearing stamen located at the base of the lip in the lady's slippers.

Stigma: the top of the pistil, the part that receives the pollen.

Style: the part of the pistil between the stigma and the ovary.

Succulent: fleshy, juicy.

Symbiosis: a mutually beneficial relationship between two dissimilar organisms.

Synsepal: the united lateral sepals found in the lady's slipper orchids and located beneath the pouch.

Tuber: in orchids, a thickened or swollen underground part of a stem.

Tubercle: a small rounded prominence or swelling, a "bump."

Viscidium: the sticky pad of the pollinarium that attaches to the pollinator.

Bibliography

Atwood, Dr. John T., Jr. "In Defense of *Cypripedium kentuckiense* C. F. Reed." *American Orchid Society Bulletin* (August 1984): 835–41.
Brooks, Maurice. *The Appalachians.* Boston: Houghton Mifflin, 1965.
Brown, Paul Martin. *Wild Orchids of the Northeastern United States: A Field Guide.* Ithaca: Cornell University Press, 1997.
Campbell, Carlos C., William F. Hutson, and Aaron J. Sharp. *Great Smoky Mountains Wildflowers.* 3rd ed. Knoxville: University of Tennessee Press, 1970.
Case, Frederick W., Jr. *Orchids of the Western Great Lakes Region.* Rev. ed. Bulletin 48. Bloomfield Hills, Mich.: Cranbrook Institute of Science, 1987.
Catling, Paul M., and Katharine B. Gregg. "Systematics of the Genus *Cleistes* in North America." *Lindleyana* 7, no. 2 (1992): 57–73.
Chester, Edward W., A. Murray Evans, Hal R. Deselm, Robert Kral, and B. Eugene Wofford. *Atlas of Tennessee Vascular Plants.* Vol. 1. Clarksville, Tenn.: Austin Peay State University, 1993.
Coleman, Ronald A. *The Wild Orchids of California.* Ithaca: Cornell University Press, 1995.
Correll, Donovan Stewart. *Native Orchids of North America, North of Mexico.* Stanford: Stanford University Press, 1978.

Duncan, Wilbur H., and Leonard E. Foote. *Wildflowers of the Southeastern United States*. Athens: University of Georgia Press, 1975.

Ettman, James K., and David R. McAdoo. *Kentucky Orchidaceae*. Morrilton, Ark.: published by the authors, 1978.

Fernald, M. L. *Gray's Manual of Botany*. 8th ed. New York: D. Van Nostrand, 1970.

Freeman, Orville L., and Michael Frome. *The National Forests of America*. New York: G. P. Putnam, 1968.

Freudenstein, John V. "Systematics of *Corallorhiza* and the Corallorhizanae." Ph.D. diss., Cornell University, 1992.

Frye, Keith. *Roadside Geology of Virginia*. Missoula, Mont.: Mountain Press, 1986.

Goerke, Heinz. *Linnaeus*. New York: Charles Scribner's Sons, 1973.

Goldman, Douglas H. "A New Species of *Calopogon* from the Midwestern United States." *Lindleyana* 10, no. 1 (1995): 37–42.

Gupton, Oscar W., and Fred C. Swope. *Wild Orchids of the Middle Atlantic States*. Knoxville: University of Tennessee Press, 1986.

Harvill, A. M., Jr., Ted R. Bradley, Charles E. Stevens, Thomas F. Wieboldt, Donna M. E. Ware, Douglas W. Ogle, Gwynn W. Ramsey, and Gary P. Fleming. *Atlas of the Virginia Flora*. 3rd ed. Burkeville: Virginia Botanical Associates, 1992.

Henry, LeRoy K., Werner E. Buker, and Dorothy L. Pearth. "Western Pennsylvania Orchids." *Castanea* 40, no. 2 (1975): 93–168.

Homoya, Michael A. *Orchids of Indiana*. Bloomington: Indiana Academy of Science, 1993.

Humphrey, Harry B. *Makers of North American Botany*. New York: Ronald Press, 1961.

Justice, William S., and C. Ritchie Bell. *Wild Flowers of North Carolina*. Chapel Hill: University of North Carolina Press, 1968.

Kanze, Edward. *The World of John Burroughs*. New York: Harry N. Abrams, 1993.

Keenan, Philip E. *Wild Orchids Across North America*. Portland, Ore.: Timber Press, 1998.

Lea, Douglass. "Secrets of a Small Pogonia." *Defenders* (Nov./Dec. 1985): 32–36.

Loughmiller, Campbell, and Lynn Loughmiller. *Texas Wildflowers*. Austin: University of Texas Press, 1994.

Luer, Carlyle A. *The Native Orchids of the United States and Canada, Excluding Florida*. Bronx: New York Botanical Garden, 1975.

Massey, A. B. *Virginia Flora*. Technical Bulletin 155. Blacksburg: Virginia Agricultural Experiment Station, 1961.

Morris, Frank, and Edward A. Eames. *Our Wild Orchids*. New York: Charles Scribner's Sons, 1929.

Niering, William A., and Nancy C. Olmstead. *The Audubon Society Field Guide to North American Wildflowers: Eastern Region*. New York: Alfred A. Knopf, 1979.

Peterson, Roger Tory, and Margaret McKenney. *A Field Guide to Wildflowers*. Boston: Houghton Mifflin, 1968.

Petrie, Dr. William. *Guide to Orchids of North America*. Blaine, Wash.: Hancock House, 1981.

Porter, Dr. Duncan M. *Rare and Endangered Vascular Plant Species in Virginia*. Blacksburg: Virginia Polytechnic Institute and State University, 1979.

Radford, Albert E., Harry E. Ahles, and C. Ritchie Bell. *Manual of the Vascular Flora of the Carolinas*. Chapel Hill: University of North Carolina Press, 1968.

Reed, C. F. "*Cypripedium kentuckiense* Reed: A New Species of Orchid in Kentucky." *Phytologia* 48, no. 5 (1981): 126–28.

———. "Additional Notes on *Cypripedium kentuckiense* Reed." *Phytologia* 50, no. 4 (1982): 286–88.

Rickett, Harold William. *The New Field Book of American Wild Flowers*. New York: G. P. Putnam's Sons, 1963.

———. *Wild Flowers of the United States: The Southeastern States*. Pt. 1. Bronx: New York Botanical Garden, McGraw-Hill, 1967.

Sheviak, Charles J. "*Cypripedium parviflorum* Salisb. I: The Small-flowered Varieties." *American Orchid Society Bulletin* (June 1994): 664–69.

Strasbaugh, P. D., and Earl L. Core. *Flora of West Virginia*. Pt. 1. Morgantown: West Virginia University, 1970.

Summers, Bill. *Missouri Orchids*. Jefferson City: Missouri Department of Conservation, 1981.

Vance, F. R., J. R. Jowsey, and J. S. McLean. *Wildflowers Across the Prairies*. Vancouver, B.C.: Greystone Books, 1984.

Weldy, Troy W., Henry T. Mlodozeniec, Lisa E. Wallace, and Martha A. Case. "The Current Status of *Cypripedium kentuckiense* (*Orchidaceae*) Including a Morphological Analysis of a Newly Discovered Population in Eastern Virginia." *Sida* 17, no. 2 (1996): 423–35.

Wharton, Mary E., and Roger W. Barbour. *A Guide to the Wildflowers and Ferns of Kentucky*. Lexington: University Press of Kentucky, 1971.

Wherry, Edgar T. *Wild Flower Guide: Northeastern and Midland United States*. New York: Doubleday, 1948.

White, Peter S. *The Flora of Great Smoky Mountains National Park: An Annotated Checklist of the Vascular Plants and a Review of Previous Floristic Work*. Atlanta: Department of the Interior, National Park Service, 1982.

Wiegand, Karl M. "A Revision of the Genus Listera." *Bulletin of the Torrey Botanical Club* 26, no. 4 (1899): 157–71.

Index

Pages shown in bold type are those for the primary species description. Photographs and range maps are included in the primary description for each species.

Adam and Eve. See
 Aplectrum hyemale
American beech. See *Fagus grandifolia*
Andrews, Albert Leroy, 175
Andrews's fringed orchid, 33, 45. See also *Platanthera ×andrewsii*
Aplectrum, 55
 A. hyemale, 55–57
Appalachian twayblade. See *Listera smallii*
Arborvitae, 110
Arethusa, 35. See also *Arethusa bulbosa*
Arethusa, 57, 64, 192, 195, 219
 A. bulbosa, 57–60
 A. ophioglossoides, 192
Augusta County, Va.: as special orchid place, 34–35

Autumn coralroot orchid. See *Corallorhiza odontorhiza*
Autumn ladies' tresses. See *Spiranthes cernua*

Bald cypress. See *Taxodium distichum*
Bayard's adder's mouth orchid. See *Malaxis bayardii*
Bentley's coralroot orchid, 37. See also *Corallorhiza bentleyi*
Blue Ridge Parkway: as special orchid place, 31–32; as threatened orchid habitat, 38–40
Bog rose, 35. See also *Arethusa bulbosa*
Bracken fern. See *Pteridium aquilinum*

Butterfly orchid. See *Platanthera psycodes*

Calopogon, 60–61
 C. tuberosus, 61–64, 194
Canada mayflower. See *Maianthemum canadense*
Cinnamon fern. See *Osmunda cinnamomea*
Cleistes, 64
 C. bifaria, 47, 64–68
 C. divaricata, 47, 64
 C. divaricata var. *bifaria*, 64
Clintonia umbellulata, 95–96
Clinton's lily. See *Clintonia umbellulata*
Club-spur orchid, 33. See also *Platanthera clavellata*

Cock's comb orchid, 36. See also *Hexalectris spicata*
Coeloglossum, 68
C. viride var. *virescens*, 68–70
C. viride var. *viride*, 68
Common plantain. See *Plantago*
Convallaria montana, 96
Corallorhiza, 70–71, 123
  C. bentleyi, 47, 71–76, 86, 211
  C. maculata, 74, 76–79
  C. maculata forma *flavida*, 46, 79, 85
  C. maculata var. *occidentalis*, 48, 78–79, 83
  C. odontorhiza, 73, 80–83, 86
  C. odontorhiza forma *flavida*, 82–83
  C. striata, 71, 72–73
  C. striata var. *involuta*, 73
  C. trifida var. *verna*, 71, 83–85
  C. wisteriana, 81, 86–90
  C. wisteriana forma *toleri*, 89–90
Cranberry Glades: as special orchid place, 30–31
Crane-fly orchid, 17, 34. See also *Tipularia discolor*
Crested coralroot orchid, 26, 36. See also *Hexalectris spicata*
Crested fringed orchid, 35, 36. See also *Platanthera cristata*
Cumberland Plateau: as special orchid place, 35–36
Cypripedium, 90–91
  C. acaule, 45, 91, 91–96, 120, 132
  C. calceolus, 47, 103
  C. candidum, 96–98
  C. daultonii, 99
  C. kentuckiense, 47, 99–103
  C. parviflorum, 47, 48, 97, 99, 103–7
  C. parviflorum var. *parviflorum*, 45, 97, 106

C. parviflorum var. *pubescens*, 106
C. reginae, 101, 107–10, 205

Daulton's lady's slipper. See *Cypripedium daultonii*
Diana fritillary butterfly. See *Speyeria diana*
Dicentra canadensis, 114
  D. cucullaria, 114
Downy rattlesnake plantain, 15, 19. See also *Goodyera pubescens*
Dragon's mouth orchid. See *Arethusa bulbosa*
Dutchman's breeches. See *Dicentra cucullaria*

Early pink azalea. See *Rhododendron periclymenoides*
Early spotted coralroot orchid. See *Corallorhiza maculata* var. *occidentalis*
Eastern prairie fringed orchid, 35. See also *Platanthera leucophaea*
Epargyreus clarus, 164
Epipactis, 111
  E. helleborine, 111–13

Fagus grandifolia, 217
Fall coralroot orchid. See *Corallorhiza odontorhiza*
Fen orchid. See *Liparis loeselii*
Fragrant ladies' tresses. See *Spiranthes odorata*
Fringed orchids. See *Platanthera*

Galearis, 113
  G. spectabilis, 113–17
Goodyer, John, 117
Goodyera, 117
  G. pubescens, 95, 118–20, 199
  G. repens var. *ophioides*, 120–22

Grass-pink orchid, 10, 31. See also *Calopogon tuberosus*
Grass-pink orchids. See *Calopogon*
Gray, Asa, 212
Great Plains ladies' tresses. See *Spiranthes magnicamporum*
Great Smoky Mountains: as special orchid place, 33–34; as threatened orchid habitat, 40
Green adder's mouth orchid, 26, 32, 33, 46, 47. See also *Malaxis unifolia*
Green fringed orchid. See *Platanthera lacera*
Green frog orchid. See *Coeloglossum viride* var. *virescens*

Habenaria, 68, 150–51
Heart-leaved twayblade, 30. See also *Listera cordata*
Helleborine orchid, 45. See also *Epipactis helleborine*
Helleborine orchids. See *Epipactis*
Helleborus orientalis, 111
Heller, A. A., 144
Henderson County, N.C.: as special orchid place, 34–35
Hexalectris, 123
  H. spicata, 123–26

Indian cucumber root. See *Medeola virginiana*
Isotria, 126
  I. medeoloides, 126–29
  I. verticillata, 95, 120, 130–32

Jones, J. I. (Bus): *Liparis* ×*jonesii* named for, 138
Jones's twayblade, 45. See also *Liparis* ×*jonesii*

Index **232**

Kalmia latifolia, 132, 147
Keenan, Philip E.: *Platanthera ×keenanii* named for, 176
Keenan's fringed orchid, 31. See also *Platanthera ×keenanii*
Kentucky lady's slipper, 36. See also *Cypripedium kentuckiense*
Kidney-leaved twayblade. See *Listera smallii*

Ladies' tresses orchids. See *Spiranthes*
Lady's slipper orchids. See *Cypripedium*
Large cranberry. See *Vaccinium macrocarpon*
Large pad-leaf orchid, 45. See also *Platanthera orbiculata* var. *macrophylla*
Large purple fringed orchid, 10, 32, 171, 173, 175, 189. See also *Platanthera grandiflora*
Large rosebud orchid. See *Cleistes divaricata*
Large whorled pogonia, 16, 37. See also *Isotria verticillata*
Large yellow lady's slipper. See *Cypripedium parviflorum* var. *pubescens*
Lenten rose. See *Helleborus orientalis*
Lesser rattlesnake plantain, 18–19, 32. See also *Goodyera repens* var. *ophioides*
Lily-leaved twayblade, 32. See also *Liparis liliifolia*
Lily of the valley. See *Convallaria montana*
Linnaeus, Carolus, 44, 103, 136
Liparis, 132–33
  L. ×*jonesii*, 138–40
  L. liliifolia, 133–36, 138, 140
  L. loeselii, 135, 136–38
Lister, Martin, 140

Listera, 133, 140
  L. cordata, 140–44, 147
  L. reniformis, 144
  L. smallii, 144–47
Little ladies' tresses. See *Spiranthes tuberosa*
Loesel, Johann, 136
Loesel's twayblade, 16, 17, 32. See also *Liparis loeselii*
Long-bracted orchid. See *Platanthera flava* var. *herbiola*

Maianthemum canadense, 133
Malaxis, 133, 147
  M. bayardii, 150
  M. unifolia, 148–50
Massie Gap: as special orchid place, 33
Mauve sleekwort. See *Liparis liliifolia*
Medeola virginiana, 96, 126
Monkey face orchid, 35–36. See also *Platanthera integrilabia*
Mountain laurel. See *Kalmia latifolia*

Nodding ladies' tresses, 18, 31, 33. See also *Spiranthes cernua*
Nodding pogonia. See *Triphora trianthophora*
Northern coralroot orchid, 30. See also *Corallorhiza trifida* var. *verna*
Northern rein orchid. See *Platanthera hyperborea*
Northern slender ladies' tresses. See *Spiranthes lacera* var. *lacera*
Northern tubercled orchid. See *Platanthera flava* var. *herbiola*

Orange fringed orchid. See *Platanthera ciliaris*
Orchis spectabilis, 113
Osmunda cinnamomea, 68
Oval ladies' tresses. See *Spiranthes ovalis* var. *erostellata*

Pad-leaf orchid, 10. See also *Platanthera orbiculata*
Phlox paniculata, 187
Pink lady's slipper, 15, 19, 32, 33, 37. See also *Cypripedium acaule*
Pink moccasin flower. See *Cypripedium acaule*
Plantago, 117
Platanthera, 55, 68, 150–51, 196, 197
  P. ×*andrewsii*, 175
  P. blephariglottis var. *blephariglottis*, 153
  P. blephariglottis var. *conspicua*, 153, 170
  P. ciliaris, 155–57, 159, 170, 188
  P. clavellata, 157–58, 181
  P. cristata, 159–60, 169
  P. flava, 48, 160–64
  P. flava var. *flava*, 162
  P. flava var. *herbiola*, 162–64
  P. grandiflora, 164–67, 171, 173, 189
  P. hyperborea, 160
  P. integra, 168–70, 170
  P. integrilabia, 153, 170, 170–72
  P. ×*keenanii*, 45, 167, 175–76
  P. lacera, 167, 172–76, 181, 188, 190, 201
  P. leucophaea, 96, 178–79, 181
  P. orbiculata, 151, 181–84
  P. orbiculata var. *macrophylla*, 182
  P. orbiculata var. *orbiculata*, 182–83
  P. peramoena, 171, 184–89
  P. praeclara, 178
  P. psycodes, 164, 173, 189–92
Pogonia, 192
  P. ophioglossoides, 192–94
Ponthieu, Henri de, 195
Ponthieva, 195
  P. racemosa, 195–97
Psychidae, 189
Pteridium aquilinum, 67

Purple fringeless orchid, 26, 29, 171. See also *Platanthera peramoena*
Puttyroot orchid, 17. See also *Aplectrum hyemale*

Rafinesque, C. S., 99, 113
Ragged fringed orchid, 26, 32. See also *Platanthera lacera*
Rattlesnake orchids. See *Goodyera*
Rattlesnake plantains. See *Goodyera*
Rhododendron catawbiense, 147
R. maximum, 147
R. periclymenoides, 96
Rosebud orchids. See *Cleistes*
Rose pogonia, 31. See also *Pogonia ophioglossoides*

Shadow witch orchid, 35. See also *Ponthieva racemosa*
Shining ladies' tresses, 26, 37. See also *Spiranthes lucida*
Short-bracted orchid. See *Platanthera flava* var. *flava*
Shortia, 27
Showy lady's slipper, 36. See also *Cypripedium reginae*
Showy orchis, 17, 18, 26. See also *Galearis spectabilis*
Silver-spotted skipper. See *Epargyreus clarus*
Slender ladies' tresses, 19, 26, 48. See also *Spiranthes lacera*
Small, John Kunkel, 144
Small purple fringed orchid, 32, 33, 164, 165, 173, 175. See also *Platanthera psycodes*
Small white lady's slipper. See *Cypripedium candidum*
Small whorled pogonia, 15, 35, 40–41, 126. See also *Isotria medeoloides*
Small yellow lady's slipper. See *Cypripedium parviflorum* var. *parviflorum*
Smaller pad-leaf orchid, 45. See also *Platanthera orbiculata* var. *orbiculata*
Smaller rosebud orchid, 36. See also *Cleistes bifaria*
Small's twayblade, 33. See also *Listera smallii*
Snake mouth orchid. See *Pogonia ophioglossoides*
Southern belle orchid. See *Platanthera peramoena*
Southern rein orchid. See *Platanthera flava*
Southern slender ladies' tresses. See *Spiranthes lacera* var. *gracilis*
Southern tubercled orchid. See *Platanthera flava* var. *flava*
Speyeria diana, 157
Spiranthes, 197–98
S. cernua, 198–99
S. grayi, 212
S. lacera, 48, 201–3
S. lacera var. *gracilis*, 202
S. lacera var. *lacera*, 202–3
S. lucida, 203–5
S. magnicamporum, 198, 206–7
S. ochroleuca, 198, 208–9
S. odorata, 198
S. ovalis var. *erostellata*, 210–11
S. tuberosa, 212–13
S. vernalis, 214–16
Spotted coralroot orchid, 31. See also *Corallorhiza maculata*
Spreading pogonia. See *Cleistes bifaria*
Spring ladies' tresses, 26, 32. See also *Spiranthes vernalis*
Squirrel corn. See *Dicentra canadensis*
Striped coralroot orchid. See *Corallorhiza striata*

Tall phlox, 29. See also *Phlox paniculata*
Taxodium distichum, 197
Three-birds orchid, 16. See also *Triphora trianthophora*
Tipularia, 216
T. discolor, 216–19
Toler, G. R. (Bobby): *Corallorhiza wisteriana* forma *toleri* named for, 90
Toler's coralroot orchid. See *Corallorhiza wisteriana* forma *toleri*
Triphora, 219
T. trianthophora, 219–22
True twayblades. See *Listera*
Tubercled orchid. See *Platanthera flava*
Twayblade orchids. See *Liparis*

Vaccinium macrocarpon, 64

Western prairie fringed orchid. See *Platanthera praeclara*
White cedar, 110
White fringed orchid, 35, 170, 171. See also *Platanthera blephariglottis* var. *conspicua*
White fringeless orchid, 35–36. See also *Platanthera integrilabia*
Whorled pogonias. See *Isotria*
Wiegand, Karl, 144
Wister's coralroot orchid, 18, 26. See also *Corallorhiza wisteriana*

Yellow-form autumn coralroot orchid. See *Corallorhiza odontorhiza* forma *flavida*
Yellow-form spotted coralroot orchid, 46. See also *Corallorhiza maculata* forma *flavida*
Yellow-form striped coralroot orchid. See *Corallorhiza striata*

Yellow-form Wister's coral-root orchid. See *Corallorhiza wisteriana* forma *toleri*
Yellow fringed orchid, 26, 32, 41–42, 159. See also *Platanthera ciliaris*
Yellow fringeless orchid, 36, 170. See also *Platanthera integra*
Yellow ladies' tresses, 31. See also *Spiranthes ochroleuca*
Yellow lady's slipper, 10, 13, 15, 34, 37, 45, 47, 48. See also *Cypripedium parviflorum*
Yellow moccasin flower. See *Cypripedium parviflorum*